Cambridge Elements ≡

Elements in the Philosophy of Science
edited by
Jacob Stegenga
University of Cambridge

NATURAL KINDS

Muhammad Ali Khalidi
City University of New York

Shaftesbury Road, Cambridge CB2 8EA, United Kingdom

One Liberty Plaza, 20th Floor, New York, NY 10006, USA

477 Williamstown Road, Port Melbourne, VIC 3207, Australia

314–321, 3rd Floor, Plot 3, Splendor Forum, Jasola District Centre,
New Delhi – 110025, India

103 Penang Road, #05–06/07, Visioncrest Commercial, Singapore 238467

Cambridge University Press is part of Cambridge University Press & Assessment,
a department of the University of Cambridge.

We share the University's mission to contribute to society through the pursuit of
education, learning and research at the highest international levels of excellence.

www.cambridge.org
Information on this title: www.cambridge.org/9781009005067

DOI: 10.1017/9781009008655

When citing this work, please include a reference to the DOI: 10.1017/9781009008655

First published 2023

A catalogue record for this publication is available from the British Library.

ISBN 978-1-009-00506-7 Paperback
ISSN 2517-7273 (online)
ISSN 2517-7265 (print)

Natural Kinds

Elements in the Philosophy of Science

DOI: 10.1017/9781009008655
First published online: August 2023

Muhammad Ali Khalidi
City University of New York
Author for correspondence: Muhammad Ali Khalidi, makhalidi@gc.cuny.edu

Abstract: Scientists cannot devise theories, construct models, propose explanations, make predictions, or even carry out observations without first classifying their subject matter. The goal of scientific taxonomy is to come up with classification schemes that conform to nature's own. Another way of putting this is that science aims to devise categories that correspond to "natural kinds." The interest in ascertaining the real kinds of things in nature is as old as philosophy itself, but it takes on a different guise when one adopts a naturalist stance in philosophy, that is, when one looks closely at scientific practice and takes it as a guide for identifying natural kinds and investigating their general features. This Element surveys existing philosophical accounts of natural kinds, defends a naturalist alternative, and applies it to case studies in a diverse set of sciences. This title is also available as Open Access on Cambridge Core.

Keywords: natural kinds, taxonomy, classification, scientific categories, naturalism

ISBNs: 9781009005067 (PB), 9781009008655 (OC)
ISSNs: 2517-7273 (online), 2517-7265 (print)

Contents

1 The Metaphysics of Kinds

The topic of natural kinds, the real divisions or groupings that exist in the world, straddles metaphysics and philosophy of science.[1] Since the underlying nature of reality has traditionally been the province of metaphysics, natural kinds have long been the stuff of philosophical theorizing, but since science is widely regarded as our best guide to the kinds of things that exist, it is increasingly clear that any discussion of natural kinds needs to be informed by the philosophy of science and, indeed, by science itself. Consequently, in this first section, I will discuss some of the metaphysical background relevant to kinds from the perspective of the philosophy of science. In Section 1.1, I will briefly consider the history of the concept of "natural kind" and elaborate the notion of naturalness at issue. Then, in Section 1.2, I will contrast realist and anti-realist views of natural kinds, and in Section 1.3, I will distinguish various forms of pluralism about natural kinds. Finally, in Section 1.4, I will discuss the relationship between kinds and properties.

1.1 What Is Natural about Natural Kinds?

The expression "natural kind" is one of those terms of art that are ubiquitous in philosophical discourse but hardly known outside of philosophical circles. Does that mean that the question of what natural kinds are – and which natural kinds there are – is of no interest to practicing scientists, just to philosophers who theorize about science? In what follows, I will try to show that that is not the case, and that scientists think about these questions, though rarely under this description. I will also try to show that they *ought* to think about them and may have something to learn from philosophical debates about natural kinds. Equally, philosophers have plenty to learn from scientific debates about the validity of constructs, the objectivity of taxonomic categories, and similar questions, which are closely related to questions about natural kinds.

  The history of the expression "natural kind" can be traced back to the mid-nineteenth century, though related issues have been posed throughout the history of philosophy. Ian Hacking (1991a, 110) has pointed out that the earliest known use of the term occurs in the work of the logician and philosopher John Venn (famous for his eponymous diagrams). Venn (1866/1888) was explicitly building on work by two illustrious predecessors, William Whewell (1840/

---

[1] In the past half-century, much of the philosophical discussion of natural kinds has been rooted in the philosophy of language, since natural kind terms have been widely thought to have a distinctive semantics that resembles the semantics of proper names. But that claim is controversial, and it is anyway not clear what implication this semantic thesis has for natural kinds themselves, so I will not delve into those discussions.

1847) and John Stuart Mill (1843/1882), both of whom were interested in scientific classification and taxonomy. But unlike Venn, neither of them used the expression "natural kind" to talk about scientific classification. Indeed, I would venture so far as to say that Venn's use of the term may have been a historical accident, since he incorrectly attributes the expression to Mill.[2]

Both Whewell and Mill were interested in "natural classification" as opposed to artificial; they were also intent on identifying the "kinds" that correspond to the taxonomic categories of a natural classification scheme. But they do not seem to have denoted them using the expression "natural kind." Why does this matter? Because the term "natural" is unfortunate in the expression "natural kind" and has led to some misleading claims and conclusions. So I will try to avoid it as far as possible, speaking simply of "kinds," or to use another expression that Mill sometimes deployed, "real kinds."

The adjective "natural" is unfortunate for at least three reasons. One is that it is multiply ambiguous, as witnessed by the fact that it has a number of different opposites: unnatural, supernatural, artificial, artifactual, social, conventional, and arbitrary, to name several. Perhaps the sense that is closest to what philosophers intend is that which opposes *arbitrary* or *conventional*, since natural kinds are supposed to be genuine, real, or principled. The second reason is related to the first, namely that it tends to lead to certain unwarranted inferences; for example, that there can be no natural kinds in the social world or in the domain of artifacts. But that would only follow if "natural" in this context was meant to be opposed to social or artifactual (or artificial). However, that does not seem to have been the intent of the originators of the expression, nor is it the conclusion of many contemporary philosophers – and if it is, it should be arrived at only after substantial argumentation, not by semantic fiat. A third problem with the epithet "natural" is that what is natural is often equated with what is normal, or what ought to be the case. But no such normative dimension attaches to kinds, since they are just there in the world; it's not that they *ought* to be there.

In what follows, I will assume that the "natural" in "natural kinds" is intended to contrast with the arbitrary (or perhaps the conventional) and that the purpose of identifying natural kinds is to isolate groupings or patterns in the world that are real as opposed to spurious. This seems consistent with the usage of Mill (1843/1882, IV.vii.4; emphasis added) who writes: "In so far as a natural classification is grounded on real Kinds, its groups are certainly not *conventional*: it is perfectly true that they do not depend upon an *arbitrary* choice of the naturalist." This position presupposes at least a moderately realist attitude

---

[2] In one use of the expression, Venn (1889/1907, 84) wrote that Mill "introduced the technical term of 'natural kinds' to express such classes as these." It is unclear whether Venn misremembered Mill's terminology or whether he deliberately modified it.

toward science and its deliverances, according to which established scientific theories are true, or approximately true, and scientific terms, including classificatory terms, successfully refer to entities in the world. (What if we're not realists about science? In the next section, we will explore the difference between realists and anti-realists in this regard.) Consequently, in this first section, I will proceed by assuming that scientific categories are a defeasible guide to the kinds that exist in the universe.

When we isolate real groupings or patterns, what we are doing is collecting things together based on their membership in kinds. Kinds have members that are really related to each other, not just arbitrarily or conventionally. A group of individuals all of which belong to the same kind is characterized by a real or objective relation, in some sense to be specified. They include the collection of atoms of uranium-238 and the collection of angiosperms, but not the collection of atoms of elements whose names start with the letter "b" (in English), or the collection of plants with pink flowers. We think of uranium-238 and angiosperms as real kinds, but not haphazard collections of things. It is customary to think of the relation that unites these items as a similarity relation and to say that members of a kind are all similar to one another (Quine 1969). But similarity does not seem to help clarify matters since it is an obscure relation, and in some sense, everything is similar to everything else. After all, atoms of elements whose names start with the letter "b" are similar in that very respect. The challenge is to say which respects of similarity pertain to kinds and which don't, or which groups or collections are kinds and which aren't. In Section 1.4, we will see that a more promising approach is to say that members of kinds share a number of properties instead of saying that they are similar to one another. In other words, what real kinds generally have and arbitrary groupings lack is a cluster of shared properties. But that proposal brings its own challenges, as we shall see.

1.2 Realism and Anti-Realism (aka Conventionalism)

It is customary to distinguish *kinds*, which belong to the world, and *categories*, which pertain to our language and theories.[3] If we do so, we can say that one aim of science is to make scientific categories correspond to kinds. (In some areas of science, this is often referred to as ensuring the *validity* of one's categories or constructs, or *construct validity*, for short.) We should be careful in stating this aim, however, since this way of talking might tempt us to think that we can directly compare categories to kinds, as one might compare a passport photo to

---

[3] Some philosophers use the term "category" roughly interchangeably with the term "kind." But according to the usage I will be adopting, a category is a concept, specifically, a classificatory concept.

its holder, or a map to a geographical region. But that would be misguided since we cannot just juxtapose our categories with our kinds in this direct fashion. There's no way to get at the kinds except by means of our language, theories, and categories, but that doesn't mean that we can't tell which categories correspond to kinds and which don't. As we shall see, realists tend to think that there are ways and means of determining whether our categories correspond to kinds and hold that one aim of science is to ensure that they do so.

There are philosophers (and scientists) who would reject the aim of making our categories correspond to kinds. They would regard such a goal as misguided. Scientific categories, they might say, merely reflect our parochial human interests. Thinking that they somehow identify the real groupings or collections in the universe is mistaken. Indeed, our categories may be mere reflections of a particular culture, ideology, or historical time period and may not be universal even among humans. To suppose that the categories that we devise "carve nature at its joints" is hubris. It would be just as blinkered for a butcher to assume that the way they carve a cow is the one true way that cleaves to the real joints of the animal, while all other carvings are bogus (see Figure 1).[4]

A complete response to this way of thinking would lead us deep into the debate about scientific realism. Suffice it to say that there are a number of sophisticated alternatives to scientific realism that have been developed by philosophers. On some anti-realist views, the categories that we devise enable us to achieve certain purposes, but they cannot be presumed to identify real divisions in nature. They may be useful tools in certain respects and satisfy our need to make sense of the world, but they should not be taken to delineate the seams of nature, so to speak. Anti-realists may be willing to privilege some

(a)   (b)

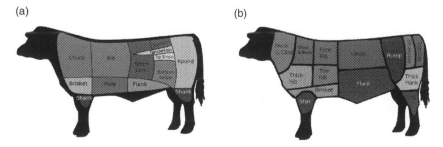

Figure 1 Diagrams showing (a) US and (b) British beef cuts. Source: Wikipedia entry "Cut of Beef."

---

[4] The overused metaphor of joint-carving can be traced back to Plato's dialogue *Phaedrus*.

categories over others, but they would balk at the claim that some categories are *really* identifying groupings or divisions in the universe. To rein in our realist tendencies, anti-realists might also cite our track record of positing bogus scientific kinds that were later discarded as we gathered more evidence and science matured.

One prominent way of pushing back on anti-realism is based on the successes of scientific theories. According to that argument, given these successes, it would be a miracle if successful scientific theories were not (approximately) true, and if their terms did not refer to real entities in the world. Even though we can't directly compare our terms and categories to the kinds, we can do so indirectly by seeing whether our theoretical categories enable us to predict future events, develop efficacious technologies, advance our understanding of phenomena, and frame satisfactory explanations. It would be a complete fluke if our scientific theories and their categories were such effective instruments yet did not identify real divisions in the world. Rather than posit a miracle, we should accept the literal truth of scientific theories and admit that their categories correspond to real kinds. This is the so-called no-miracles argument for scientific realism (Putnam 1975).[5] In this context, the argument would consist in saying that the categories that feature in scientific theories appear to be highly effective in enabling us to explain, predict, control, manipulate, and otherwise interact with the world. In fact, they seem to be getting more successful as science becomes more sophisticated, and as it discards some useless categories and devises a plethora of new ones in a range of emerging disciplines and subdisciplines. The success of scientific categories is surely some indication that they are engaging with the world, as it were; that our classificatory gears are making contact with the cogs of the universe itself rather than spinning in a void. Thus, the claim that successful scientific categories often identify kinds is warranted, and science should aim to come up with categories that do so.

In response to this realist stance, there are "semi-realist" or "quasi-realist" positions that concede the successes of scientific theories, along with their classificatory categories and taxonomic schemes, yet claim that these categories are not simply reflections of the divisions in nature. They could be a result of some combination of our own human interests or predilections and objective features of reality. Massimi (2014, 434) articulates a "mild" realist position according to which natural kinds depend "on our conditions of possibility of having a comprehensible experience of nature" and are hence dependent on our human capacities rather than mere reflections of the kinds that exist in the world.

---

[5] For a more recent version of the success-to-truth argument, see Kitcher (2001, 16–28), though it does not directly invoke categories and kinds.

A semi-realist position may also be supported by observing that in many scientific domains, there are often a number of slightly different ways of drawing and redrawing category boundaries, none of which is privileged. If that is the case, we should accord some role to convention, convenience, or other human-centered factors in delimiting our scientific categories. This would make scientific categorization at least partly a conventional matter and would also pose a challenge to a pure realist position (cf. Reydon 2016; Ludwig 2018; Boyd 2021).

Realists may counter by agreeing that scientific categories serve our interests, provided those interests are epistemic and consist in offering an accurate account of the kinds that exist in the world. After all, one of the main purposes of science is to explain and understand the world, which means (in part) ascertaining what kinds of things exist. Hence, they would say that it is up to anti-realists to show that there are *other* interests at play in delimiting the boundaries of scientific categories, and that these interests are not, whether directly or indirectly, in the service of accurately identifying the kinds of things there are in the world. Some authors tend to jump too quickly from the claim that categorization is (1) interest-relative or (2) discipline-specific to the conclusion that (3) natural kinds are conventional or not real. For example, Boyd (2021, 2864) writes that approaches to natural kinds that hold that natural kinds "are discipline specific and thus mind and interest dependent … can be challenged as non-realist for that reason." But both moves are unwarranted if we bear in mind that (1*) an interest in discerning groupings in nature can reveal real kinds, and (2*) that scientific disciplines are dedicated to discovering the divisions in their respective domains. There is nothing inherently anti-realist or conventionalist in the claim that categorization is interest-relative or discipline-specific.[6] (In Section 3.4, we will revisit this semi- or quasi-realist position in the context of discussing kinds that are mind-dependent in certain ways.)

There is a different variety of realism when it comes to kinds, which tackles the metaphysical status of kinds.[7] Even if we agree with the scientific realist that there are objective divisions in nature, that one aim of science is to identify them, and that we have reason to think that science has identified

---

[6] Magnus (2018) distinguishes the "simpliciter view" of kinds, according to which a kind is just a kind, full stop, from the "relation view," according to which a kind is always a kind for a certain domain. But he also writes: "The domain–relative conception of natural kinds allows us to say that the natural kinds discovered by science are genuine, objective features of the domains which scientists have investigated" (2018, 10). This last point accords with the position defended here and, as far as I can tell, it would seem to collapse the distinction between the two views.

[7] As we will see in Section 2, many theories of kinds give a metaphysical account of them in terms of such things as essences, mechanisms, or causal structures. The question here is whether there is some *additional* metaphysical entity that corresponds to the kind itself (for further discussion, see Hawley and Bird (2011)).

many of them, we can ask about the ultimate nature of these divisions. Do the groupings that we identify consist in some additional metaphysical entities, or are they nothing more than the sum of their members? This is the traditional debate between metaphysical realists and nominalists, or to avoid confusion with *scientific* realism, between universalists and nominalists. Universalists would say that kinds are entities in their own right and that their members *belong* to them or *instantiate* them. For example, there is such a thing as the element beryllium over and above all the atoms of beryllium that there ever were and ever will be. This entity, the kind itself, can be thought of as a transcendental entity, something like a Platonic form in which its members participate (or which its members instantiate), or else it could be considered immanent in all its members (Armstrong 1989). Nominalists, by contrast, think that though there is an objective relationship between all atoms of beryllium, positing an additional entity is ontologically profligate and super-fluous. This is a debate that I will be setting aside in what follows. Suffice it to say that the universalist–nominalist debate about kinds should not be confused with the realist–conventionalist one that was discussed earlier in this section. Nominalists can also be scientific realists, since one could affirm a real, nonarbitrary, and nonconventional relationship between members of a kind, yet deny that this corresponds to some additional metaphysical entity. It is not as clear that universalists can be anti-realists, for if scientific categories do not identify real divisions in the world, it would be odd to say that these divisions pack some further metaphysical punch.

1.3 Pluralism

Some views of kinds may be considered sparse or parsimonious, others extrava-gant or profligate. If the bar is lowered for what it is for something to be a kind, then we would let in a relatively large number of kinds, whereas if it is raised, we would admit relatively few. Pluralists about kinds are of the former persua-sion. Pluralism about kinds is sometimes thought to threaten realism. After all, if anything goes, then we're in danger of letting in the arbitrary kinds mentioned in previous sections (i.e. atoms of elements whose names start with the letter "b," plants with pink flowers). Thus, pluralists about kinds are sometimes charged with anti-realism, since their account of kinds is so extravagant. Before looking into this charge, it is worth mentioning that there are at least two kinds of pluralism about kinds, which are not always clearly distinguished.

The first type of pluralism says simply that there is a multitude of kinds in the world. It may have once seemed that the kinds were a privileged set, consisting, say, of kinds of elementary particles, chemical elements, chemical compounds, and perhaps biological species.[8] But if we take science as our (defeasible) guide to the kinds there are in the world, then clearly we would have to admit more kinds, since science currently posits many more theoretical categories, each of which could be said to be associated with a kind. As science progresses, its categories multiply, and the candidates for kinds proliferate. One of the striking features of science in the past century or so is the frequent emergence of various disciplines and subdisciplines, each with its own proprietary categorization schemes and taxonomic systems. New scientific categories are proposed on a regular basis, and though some are soon left by the wayside, being deployed in only one or a few scientific hypotheses that are later discarded, others prove useful and become entrenched in scientific practice. Scientific training in the twenty-first century consists largely in mastering a vocabulary of taxonomic categories that are wielded by experts to communicate with other experts. This raises the question as to which of them correspond to actual kinds and which are mere artifacts or convenient crutches. Assuming we don't adopt a blanket anti-realist position and consider all categories to be convenient fictions, we need to take a stance on how to decide which categories should be admitted into our ontology.

Now if we have a principled standard or criterion that kinds must satisfy, we could just admit all those that do so. But what if too many do? It's not clear what would constitute a surplus of kinds, or even on what grounds we would draw the line. In fact, given that there seems to be no principled upper limit to be set on the number of kinds in the universe, it's not even clear how to distinguish pluralists from minimalists. We could reasonably claim that anyone who only admits the kinds of elementary particles (e.g. up-quarks, electrons, neutrinos) can be considered a minimalist. But what about someone who also admits kinds of elementary atoms (e.g. hydrogen, helium, beryllium)? Or someone who admits those as well as kinds of chemical compound? Biological species? There does not seem to be a clear or principled way of demarcating pluralism from minimalism. Still, we can say roughly that while minimalists cleave to a traditional view that kinds can be restricted to a relatively small set consisting of such paradigmatic kinds as elementary particles, chemical elements, and chemical compounds (though even these amount to millions of kinds), pluralists posit many more kinds, perhaps roughly as many kinds as there are (successful, nonredundant) theoretical categories in a successful science.

Another type of pluralism can be distinguished from the first (although they are certainly compatible and the second may even imply the first, on certain

---

[8] Even this would leave us with millions of kinds. There are an estimated 8.7 million extant biological species. In addition, there are over 13 million stable chemical compounds with up to eleven atoms of either carbon, nitrogen, oxygen, or fluorine, according to one estimate (www .science.org/content/blog-post/just-many-compounds-we-talking).

reasonable assumptions). I mentioned that some philosophers raise the bar and others lower it when it comes to kinds. One could also introduce more than one bar. In other words, one view of kinds would have it that there are different criteria as to what counts as a kind. (In Section 2, we will look at such criteria in more detail.) Pluralism of this sort raises the question as to whether the different criteria all identify kinds or whether they identify different kinds of kinds. That is, we could posit $kind_1$, $kind_2$, $kind_3$, and so on, depending on which criteria they satisfy (see Ludwig 2018 for a proposal along these lines). This could lead us to conclude that there are no natural or real kinds and that there are a number of different types of groupings in nature. This is the view of some recent writers on natural kinds, who have denied that natural kind is itself a natural kind (Dupré 2002; Hacking 2007). Alternatively, we might consider one of them to correspond to real kinds and the others as constituting other types of groups, designating them with other terms to avoid confusion.

Does pluralism of either variety threaten realism about kinds? As we have just seen, second-order pluralism need not imply anti-realism. And as for first-order pluralism, just because there are many kinds, perhaps many more than philosophers once imagined, that doesn't mean that they aren't all real (see Ereshefsky 2001, 45). Still, there may be a lingering feeling among minimalists that kinds were supposed to be an elite bunch, few and far between, not "as plentiful as blackberries." Pluralists might respond by saying that they don't multiply kinds without necessity, but rather that the universe has just turned out to be more complex than we once imagined it to be. Consider something like the periodic table, perhaps everyone's favorite taxonomy of kinds. At first glance, it seems to classify atoms into around 100 or so kinds, each distinct from the others. That seems like a manageable number. But if we examine it more closely, we may find grounds to distinguish different isotopes of the same element, each of which constitutes a distinct subordinate kind (e.g. *uranium-238* and *uranium-235*). This would increase the number of kinds severalfold. We may also want to identify certain superordinate kinds such as *noble gases* and *alkali metals*. That would add yet more kinds, perhaps running into the thousands. Other domains are even more promiscuous in terms of the categories that we can identify. In biology, the preeminent kinds are species (e.g. *Drosophila melanogaster*, *Panthera tigris*), but biologists also identify higher taxa, like genera, families, and orders, not to mention superfamilies, subgenera, and so on. There are also many other ways of classifying organisms, based on their behavior (e.g. *nocturnal*, *diurnal*), the ecological niches they occupy (e.g. *predator*, *prey*), and life stage (e.g. *larva*, *pupa*, *imago*), among other features. Thus, it seems as though nature is full of joints and seams, and humans may have their pick of which ones to single out, label, and deploy in their theorizing.

However, if it turns out that in many domains there are numerous similar ways of carving up a domain, all in the same vicinity, and we might do just as well to single out some ways rather than others, then it appears that which ones we happen to single out may be a matter of convention. This may give a boost to the conventionalist or anti-realist position, which was mentioned in Section 1.2, as opposed to the realist one. But is it anti-realist to say that there is a plenitude of kinds, many more than we might want to keep track of, so it is up to us which ones we do in fact label and theorize about? It seems that a pluralist realist could maintain that our categories pick out real kinds, even though there are many more kinds to be identified in the world than we are able to survey. As Magnus (2012, 7) puts it: "There are so many joints in the world that we could not possibly carve it up along all of them." Given that this is the case, convenience or convention may lead us to focus on some rather than others, but these factors need not play a role in determining which kinds there are.

Whether we are realists or conventionalists, universalists or nominalists, minimalists or pluralists, it should be mentioned that the particular entities that belong to kinds need not be concrete spatiotemporal objects. There could also be kinds of process, kinds of event, kinds of capacity, and so on. In speaking of "things" or "entities" so far, I didn't mean to rule out the possibility of these other kinds of items as well. This leads to a third type of pluralism about kinds, a pluralism about higher-order ontological categories. Just as there are kinds of objects like *atoms*, there could be kinds of events, such as *ionization*, and just as there are kinds of objects like *plants*, there could be kinds of process, such as *photosynthesis*. Ionization is a repeatable event whose members are individual ionization events that occur at particular points in space and time, as when a specific atom loses an electron and becomes positively charged. Similarly, photosynthesis is a repeatable process whose members are individual processes of photosynthesis that transpire in a particular plant over a particular period of time. In addition to objects, events, and processes, do kinds also apply to other ontological categories, such as capacities, mechanisms, dispositions, systems, and so on? Scientists studying some domains also have occasion to posit these broader ontological categories, and unless one can show that some of them are reducible to others (for example, that processes are just sequences of spatiotemporally contiguous events, or mechanisms are combinations of entities and processes), there would seem to be no impediment to allowing kinds of these other types, in addition to kinds of object. This is grist for the pluralist's mill and provides further reasons for positing a plenitude of kinds in the universe.

A final question when it comes to pluralism pertains to scientific categories that do not seem, at least at first sight, to be aimed at singling out kinds. The scientific categories that lend themselves to kind-thinking are the ones that apply categorically to individuals, those that have individual entities as members, whether kinds of object, event, process, and so on. But what about other theoretical categories in science like magnitudes or quantities? Are *mass* and *charge* kinds? Or would the kind be a determinate quantity, such as *mass of one kilogram*, whose members include the laptop I'm writing on and the pineapple I bought last week? There doesn't seem to be a principled reason for denying that *mass*, or perhaps *massive object*, is a candidate for being a kind. The fact that it is also a magnitude or quantity should not pose an obstacle. After all, *mass* (or *massive object*) does apply categorically to objects even though it is also a magnitude: quarks and electrons belong to this kind, whereas electric fields and ionization events do not. In this and other cases, in addition to there being determinates of this determinable kind, there may also be subordinate kinds that fall under this superordinate kind. To illustrate, *electron* is a subordinate kind of the kind *massive object*, whereas *mass of one kilogram* is a determinate of that kind. This is just to say that such categories could *potentially* correspond to kinds, not that they *actually* do so. In fact, (having a) *mass of one kilogram* is the type of property that is often cited by philosophers as not corresponding to a kind precisely because there isn't anything in common to all objects that possess that property, apart, of course, from having a mass of one kilogram. This provides a contrast with the property of being a *massive object*, all of whose members do seem to have other properties in common too, such as the property of exerting a gravitational force on other massive objects. If that case can be made, then the determinable category (*mass*) would correspond to a kind, whereas the determinate category (*mass of one kilogram*) would not.[9] Moreover, it may be that the superordinate category (e.g. *mass*) is a kind and some of its subordinate categories are (e.g. *electron*), though others are not (e.g. *barbell*) (see Figure 2). Does this mean that every scientific category might potentially belong to a kind? The upshot of this discussion is just that there is no reason to deny that scientific categories correspond to kinds based merely on the fact that they are magnitudes, quantities, or determinables. Whether they do so will depend on other considerations, which we will look at more closely in Section 2.

---

[9] Most determinate properties do not seem to correspond to kinds, for example *having a temperature of 0° Celsius*, or *having a velocity of 1 meter per second*, or *having an electrical resistance of 1 ohm*.

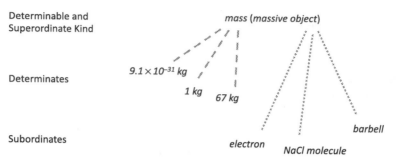

Figure 2 Diagram showing the different relations between the determinable and superordinate kind *mass* (or *massive object)* to its determinates and subordinates, some of which may be kinds in their own right, while others may not be (dashed line indicates determinable–determinate relation and dotted line indicates superordinate–subordinate relation).

1.4 Properties and Kinds

Most writers about kinds think that kinds are associated with properties. For some, kinds are nothing over and above properties or collections of properties. For others, a kind is not identical with a set of properties, though it is closely associated with a set of properties. (For example, the kind *beryllium* is identified or associated with properties such as having a certain atomic number, mass number, ionization energy, atomic radius, melting point, and so on.) Either way, we can say that members of a kind are not just similar, they share properties, or more precisely (in the limiting case), have identical properties. This is less problematic than relying on the vague notion of similarity, as mentioned in Section 1.2, and also seems like a satisfying way to situate kinds in relation to more familiar ontological categories. In Section 2, we will reexamine the relationship between kinds and properties, but for now the relation of sharing properties, in whole or in part, seems like an advance on similarity.

But linking kinds to properties in this way just leads to the question: which properties? Not arbitrary properties, like belonging to an element whose name starts with the letter "b," but genuine ones like having atomic number 4. Does this mean that the question of natural kinds is the same as or reducible to the question about "natural properties" (Lewis 1983)? By associating kinds with properties, have we merely replaced the question about natural (or real) kinds with a question about natural (or real) properties? Not quite, for a few reasons. First, the class of natural properties seems broader than the class of natural kinds. If we take science as a defeasible guide to which properties and kinds exist, it is clear that there are some properties that are used in science, such as some of the determinate properties mentioned in the previous section (e.g. *having a mass of one kilogram*) that don't seem to correspond to kinds.

Second, it is widely held that what distinguishes kinds from properties – and what led philosophers to identify them in the first place – is that they correspond to *sets* or *clusters* of properties. Though we may sometimes identify them by their central property (e.g. the kind *beryllium* can be identified with the property of having atomic number 4), that property tends to be associated with a number of others. If kinds are associated with properties, those properties are "sticky" and cluster together with other properties.[10] That seems to be a key distinguishing feature of kinds. Third, the very fact that some properties cluster together is a sign of their "naturalness," or at least their nonspuriousness. In other words, one indication that a property is natural would seem to be that it clusters with other properties. Of course, as we've seen, many determinate properties don't seem to cluster, but other determinates of the same determinables do. So the question of which properties are real or natural is not necessarily antecedent to the question of which kinds are. Having said all that, questions about natural properties and natural kinds would seem to be interrelated. This is especially so, since it is possible to construe at least some kinds themselves as properties. For example, we talk of the kind *electron*, but we can also talk about the property of *being an electron*. To be sure, the property of being an electron is equivalent to the property of being a particle with a certain mass, charge, and spin, so it is really a collection of properties rather than a single property. Still, it's sometimes just a matter of convention whether we identify the kind with one property or the whole cluster, and it's sometimes tricky to say how to individuate properties, so the distinction between kinds and properties does not seem to run very deep (but see Bird 2018 for an opposing view). To summarize, kinds can be identified with clusters of properties, or properties that cluster with other properties, and these clusters are associated not just conventionally or arbitrarily. As we shall see in the next section, they are thought to hold together in some way.

Finally, one could pose a question about the properties themselves that cluster in kinds. What makes us so sure that they pertain to the universe itself rather than to our parochial interests? Famously, Goodman (1955/1983) proposed a perverse predicate "grue" that applies to all things examined before some future time, *t*, if and only if they are green and to other things if and only if they are blue. All the evidence (collected before *t*) that confirms the empirical generalization that all emeralds are green also confirms the generalization that all emeralds are grue. Though the property *grue* seems artificial, how can we rule it out, along with other such perverse properties? Perhaps we could get *grue* to cluster with other properties by gerrymandering everything else, for instance

---

[10] Chakravartty (2007) proposes that such properties exhibit "sociability" and Slater (2015) that they evince "cliquishness," but I have put it in terms of stickiness because it's less anthropomorphic.

by defining the predicate "bleen," which applies to all things examined before some future time, t, if and only if they are blue and to other things if and only if they are green. For all we know, our science might as well have been built on an entirely different basis, and the properties we have chosen just reflect our human idiosyncracies, such as the fact that we don't index colors to times in this way. There have been a vast number of philosophical responses to Goodman's challenge, and it's not possible to survey or even sample them here. But this may be the kind of fundamental metaphysical question that does not have a satisfactory answer. I said earlier that the fact that properties cluster together seems to be one indication that they're not arbitrary – but it may be that we could get grue-like properties to cluster if we adjusted them in compensatory ways. So it may not be possible to rule out the existence of an entire set of eccentric properties that cluster in the same way that ours do and could serve as the basis for science. The question of natural kinds seems more tractable: what undergirds the clustering of properties, at least the ones that we're familiar with? That will be the focus of the next section.

2 Theories of Natural Kinds

In Section 1, we tackled the metaphysics of kinds; it's time to figure out what criteria a grouping or collection needs to satisfy to be a kind, and hence which kinds there are in the world. How can we tell which of our categories genuinely identify kinds and which are merely haphazard collections of individual entities? You might think that this is not for philosophers to say; it is scientists who conclude that *oxygen* is a kind of chemical element but *phlogiston* is not, or that *schizophrenia* is a kind of psychiatric disorder but *hysteria* is not. Still, are there any general criteria that we can identify that kinds must satisfy? In what follows, I will assume an attitude of philosophical naturalism, which takes science as a defeasible guide to the kinds of things there are in the world. This does not mean simply accepting whatever scientists say or whatever scientific theories tell us are the kinds of things in the world. Scientists sometimes make use of categories that are redundant or incidental to the real work of science. Scientists have also made mistakes in the past and introduced erroneous classifications, which have been completely discarded. They also sometimes lump phenomena into one category when they should have split them into two or more, or split them when they should have lumped them. That means that philosophers should tread with caution when it comes to the taxonomic pronouncements of science. But more important than the specific taxonomic categories or classification schemes are the principles of categorization that are either explicitly or implicitly adopted in the sciences. These are likely to be more enduring than the categories themselves and should shed some light on the underlying basis for classifying things into categories. After further explicating the naturalist stance when it comes to kinds in Section 2.1, I will present the essentialist approach to kinds in Section 2.2. After that, I will discuss a prominent alternative, the "homeostatic property cluster theory" in Section 2.3, and finally I will defend an account of kinds that I take to be more in keeping with scientific practice, the "simple causal theory" in Section 2.4.

2.1 Science as a Guide to Kinds

The most widely accepted condition that bona fide scientific categories, terms, or predicates must satisfy is that they be *projectible* (Goodman 1955/1983). A category or predicate is projectible when it figures nongratuitously in an inductive inference. More precisely, we can say that predicate P is projectible relative to predicate Q if and only if we can draw a legitimate inductive inference from x is P to x is Q, where the predicates stand for properties or natural kinds and x denotes a particular, whether an object, a specific event, an individual process, and so on (cf. Khalidi 2018, 1380–1381). It should be clear that the paradigm cases of kind categories satisfy this condition while those scientific categories that do not correspond to kinds do not. For example, we can project from *x is beryllium* to *x has a melting point of 1560 K*, and we can generalize this to: *For all x, if x is beryllium, then x has a melting point of 1560 K*. By contrast, we cannot project from *x has a mass of 1 kg* or *x has pink flowers* to any other nontrivial consequence of these claims (where trivial claims include such statements as *x has a mass of less than 2 kg*, or *x has flowers*, respectively). There may be other scientific terms or categories that do not correspond to kinds, namely those that are incidental to the work of science (e.g. *beaker, lab bench, sample, participant*), or those that are abandoned in due course (e.g. *phlogiston, hysteria*), but these are either not projectible in the first place or are discovered not to be projectible after further consideration.

In Section 1, we established that natural kinds are associated with sets of properties and that these properties cluster together or are "sticky." This basic datum is mirrored by the fact that natural kind categories or predicates are projectible. In fact, we can say that natural kind categories are projectible precisely because the properties that they denote generally cluster together and are coinstantiated in particular entities (whether objects, events, processes, and so on). That is why we can project from the presence of some properties to the presence of others. These clusters of co-occurring properties need not be replicated exactly on each occasion, which is why they can be thought of as a loose cluster of properties rather than a fixed set. However, there seems to be widespread agreement among both scientists and philosophers that for two particulars to be members of the same kind, they must share at least some properties. This means that kinds can be identified with clusters of properties, but not mere *disjunctions* of properties. If two members of an alleged kind fail to share any properties, there does not seem to be any reason to consider them to belong to the same kind. Of course, this leads to a question as to *why* the properties associated with natural kinds cluster together. A philosophical account of natural kinds needs to be able to answer this question.

Naturalism does not mean outsourcing philosophical analysis to the sciences, nor does it mean abdicating responsibility for philosophical argument. There is still a fair amount of work to be done in trying to determine the nature of real kinds. First, the underlying classificatory principles are not always explicit in the sciences and often have to be inferred from scientific practice. Various patterns and commonalities can be obscured by the differing methods and theoretical approaches adopted by different sciences and may only be ascertained after some analysis and scrutiny, which is something that philosophers are often better placed to do than scientists themselves.[11] Second, there is a need to take a bird's-eye view of the sciences and compare classificatory practices among them in order to discern what general principles apply across the sciences, if any. That is something that working scientists very seldom do, steeped as they are in their own disciplines or subdisciplines. Third, some scientific classification schemes are adopted for the sake of convenience or based on incomplete information about a domain, perhaps because the contours and divisions in that domain are difficult to ascertain. In such cases, some philosophical work may be needed to decide whether a scheme is merely provisional or expedient, or whether it is supposed to constitute a stable and settled taxonomy of that domain. If it's the former, then the taxonomic scheme in question may not be taken as a guide to real kinds.

At this point, one might ask: Why not just say that there are some categories that play an important role in inductive inference in science, feature in scientific laws and generalizations, figure in explanation and prediction, support manipulation and technological innovation, and leave it at that? Why not just affirm that some categories are projectible and have an important epistemic role to play in science, while others do not? Why the further need to say that such categories correspond to kinds? The problem with such an "epistemology only" account is

---

[11] It might be objected here that scientists are well aware of the principles that underlie their classifications, such as the various ways that biologists classify species. But, although taxonomists are usually clear on how to classify and on what basis, they aren't always explicit when it comes to the underlying principles, whether they are based on intrinsic or extrinsic properties, synchronic or diachronic, or whether they combine a number of different types of properties. The metaphysical niceties are not always brought to the fore in scientific classification.

that it just raises the metaphysical question: what is it about these categories that makes them so, and are they underwritten by certain common features in reality (cf. Lemeire 2021)? This is partly why a naturalistic approach is needed in order to take a close look at a wide range of categories in science and try to determine what (if anything) the corresponding kinds have in common. In the rest of this section, I will explicate some of the most influential accounts that have been offered of the nature of natural kinds and then will propose an alternative that builds on those accounts but purports to take a more naturalist approach than existing accounts.

2.2 Essentialist Theories

The most prevalent view of natural kinds among philosophers for the past several decades has been the essentialist one. This view says that what distinguishes natural kinds and sets them apart from other groupings in nature is that they correspond to sets of essential properties. Though there's no universal agreement on which properties are the essential ones, the following criteria are often mentioned as central to the essentialist view:

(E1) Essential properties are both necessary and sufficient for membership in the kind.

(E2) Essential properties are modally necessary to the kind (which is often glossed as: associated with the kind in every possible world in which the kind exists).

(E3) Essential properties are intrinsic to the kind rather than extrinsic or relational.

(E4) Essential properties are microstructural.

Essentialists think that for any natural kind K_1, there is a set of properties $\{P_1, P_2, \ldots, P_n\}$ such that (E1)–(E4) hold. These conditions are listed roughly in order of importance to essentialism, with (E1) and (E2) being more widely held and generally considered more important than (E3) and (E4). Natural kinds, on this view, do not just correspond to any old cluster of properties but clusters of properties that satisfy (at least some of) these conditions.[12]

---

[12] The essentialist account of natural kinds is often supported by certain claims in the philosophy of language, in particular the claim that the terms denoting natural kinds are *rigid designators* (Kripke 1972/1980). But many philosophers have recently become convinced that this claim is problematic and anyway cannot be used to support a metaphysical view of kinds. For a recent discussion, see Crane (2021), who concludes that the assumption that natural kind terms are rigid designators should be rejected.

The essentialist account of natural kinds would seem to apply well to some of the most widely accepted kinds, such as elementary particles and chemical elements. To take just one example, the kind *electron* is associated with three properties, a certain mass, charge, and spin. In accordance with (E1), every electron has these properties and anything that has these properties is an electron. Possession of these properties is what it is to be an electron. Moreover, in line with (E2), essentialists would say that this is not a contingent matter, which just happens to hold, but rather a matter of necessity. Nothing *could* be an electron unless it had these properties, and if anything has these properties then it *must* be an electron. To put it in a familiar idiom, these properties determine what it is to be an electron not just in the actual world, but in every possible world. Some essentialists also make an additional necessity claim: membership in a natural kind is necessary to the individual members of natural kinds. That means that each individual electron is necessarily an electron and could not possibly be a proton, say, or an elephant or a toothbrush, for that matter. (For the relationship between essentialism about kinds and essentialism about kind membership, see Khalidi 2009.) This also sets natural kinds apart from arbitrary kinds, since membership in arbitrary categories is thought not to be necessary to the identity of the individuals. An atom of beryllium is necessarily a beryllium atom and could not have been a molecule of water or a postage stamp. But it just happens to belong to the category of atoms of elements whose names start with the letter "b" and might not have belonged to that category had we chosen a different name for that element. The kinds that I have mentioned so far, kinds of elementary particles and chemical elements, also seem to satisfy (E3) and (E4). The properties associated with electrons are intrinsic[13] and microstructural, as are (plausibly) those associated with chemical elements.

When it comes to other widely accepted natural kinds, such as biological species or other biological categories, it is less clear whether they satisfy the essentialist criteria. Some philosophers of biology would say that species are in violation of (E1), since there is no set of properties that is common to all and only organisms that are members of a given biological species, say *Drosophila melanogaster*. Others would say that the only properties necessary and sufficient for membership in a species are historical or etiological ones, which have to do with descent from a certain historical lineage. The properties in question are not intrinsic but extrinsic or relational. This means that biological species would satisfy (E1), but only at the expense of violating (E3) and (E4). Hence,

---

[13] But elementary particles may also have some fundamental *extrinsic* or *relational* properties, such as the property of electrons (and other fermions) that they can't share the same quantum state (as dictated by the Pauli exclusion principle).

some essentialists reject (E3) and embrace origin essentialism for biological species (e.g. Griffiths 1999), while other essentialists insist on (E3) and deny that biological species are natural kinds (e.g. Ellis 2001).

What holds for biological species also holds for many other candidates for kinds in the "special sciences" (which is a misleading term used in philosophy mainly to indicate all sciences apart from fundamental physics and perhaps chemistry). It is widely recognized that essentialism, as defined by conditions (E1)–(E4), applies at best to a small portion of candidates for kinds in the sciences, such as elementary particles, chemical elements, and chemical compounds. That means that essentialism automatically rules out a huge swath of categories in the sciences as potential candidates for kinds. This might be thought to be an advantage of essentialism, at least if one is not a pluralist about natural kinds (see Section 1.3), but it would imply that most branches of science are not in the business of identifying natural kinds and that the vast majority of categories deployed in science do not pick out natural kinds. That is just because many scientific categories do not correspond to a set of necessary and sufficient properties but rather to a loose cluster of properties. Moreover, the properties involved are frequently not intrinsic or microstructural. Physics and chemistry are different from the special sciences in certain respects, such as the strictness of their laws or generalizations and the relatively limited number of terms that are involved in them. But it is not clear that natural kinds should be restricted to them. Moreover, even when it comes to the physical sciences, there would seem to be many categories that are in violation of one or more of the essentialist conditions.

A different problem with essentialism has to do with the status of the criteria (E1)–(E4) and the justification for them. They have a certain intuitive appeal, to be sure, and some of them are backed up by an authoritative lineage in the history of philosophy. When it comes to (E1), the search for necessary and sufficient conditions is, of course, a feature of much traditional philosophizing, and the distinction between what is necessarily the case and what is merely contingent also has an illustrious past. But as we come to discover a great variety of kinds in the world, it is becoming clearer that the universe may not come neatly packaged into sets of properties that are singly necessary and jointly sufficient to determine each kind. As for (E2), one of the problems with positing modal necessity is the difficulty of ascertaining what would count as evidence to support it. Scientists make counterfactual claims all the time and reason about what would happen (or would have happened) if circumstances were (or had been) different, but these speculations hardly ever seem to involve modal claims about kindhood. More to the point, if such claims *are* made, it does not seem to matter whether scientists say that (a) some kind

K might have had some other property P_2 rather than P_1, or (b) some other kind K^* might have existed that had property P_2 rather than P_1. For example, scientists might ask what would have been the case had protons had negative charge, but this seems equivalent to asking what would have been the case had there been some other kind of particle, call it *schomoton*, which is just like the kind *proton*, but with negative charge instead of positive. When it comes to (E3), there is some intuitive appeal to saying that the properties that really matter or are genuine are the intrinsic rather than the relational ones. But it is clear that a plethora of relational properties or relations play important roles in many areas of science, and many of them pertain to the physical or natural sciences, so intrinsic properties need not be privileged over relational properties. Finally, (E4) betrays a commitment to reductionism in giving priority to microstructure over macrostructure, but many philosophers no longer consider microproperties more important than macroproperties in determining the identity of a kind. In particular, if one allows that many kinds are *multiply realized* (see e.g. Fodor 1974), then their identity is not given by their microproperties but rather their macroproperties. (For more on such kinds, see Section 3.2.)

If one takes a naturalist approach, which holds that science is our best guide to natural or real kinds, then one would be dismissing the vast majority of scientific categories in endorsing an essentialist position. Moreover, one would be doing so on the say-so of a philosophical theory based largely on tradition and intuition. This should give us pause and lead us to consider other accounts of real kinds, particularly since essentialism does not offer an obvious answer to the "stickiness" question. Though essentialism places conditions on the nature of the properties associated with kinds, it does not seem to tell us what holds the properties together. In recent work, Tahko (2021, 61) attempts to remedy this by proposing that the kind itself unifies the properties. But there seems to be room for more work to be done on explicating the unification relationship within an essentialist framework.

2.3 Homeostatic Property Cluster Theory

Partly in response to the shortcomings of the essentialist account, many philosophers of science have come to embrace an account first proposed by Boyd (1989), known as the "homeostatic property cluster" (HPC) account of natural kinds. One of the main features of this account is the denial of condition (E1) in the essentialist account. As noted earlier, many categories in science correspond to a loose cluster of properties rather than a strict set. Membership in a kind is not a matter of possessing all and only a certain requisite set, but can be a matter of possessing a certain number of properties in the set. Moreover, there may be no hard-and-fast rule on how many properties or which ones have to be present for something to be a member of the kind in question. This means that members of a kind do not all share a certain set of essential properties but that their properties may only partially overlap or loosely cluster. In some scientific contexts, this is known as a *polythetic* as opposed to a *monothetic* grouping of properties. However, a crucial aspect of the HPC account is not the looseness of the cluster but the idea that the properties in the cluster are held in equilibrium or homeostasis by a causal mechanism. The causal mechanism regularly and reliably gives rise to this set of properties, but in some instances it might not, and there may be no saying in advance which of the properties will be manifested in any particular member of the kind. Still, the homeostatic mechanism is what underlies the kind.[14]

The main features of the HPC account are that natural kinds consist of (i) a mechanism that (ii) keeps the cluster of properties associated with the kind in homeostasis. There are a couple of things that need to be elaborated, namely the nature of mechanisms and the way in which mechanisms keep a cluster of properties in equilibrium. Since Boyd first proposed his account, there has been an increased interest in mechanisms in the philosophy of science. A number of philosophers think that mechanisms are central to the ontology of various sciences, particularly the biological sciences. On one very influential account, mechanisms are considered to be "entities and activities organized such that they are productive of regular changes from start or set-up to finish or termination conditions" (Machamer, Darden, and Craver 2000, 3). On another, they are described as "an organized set of parts that perform different operations which are orchestrated so as to realize in the appropriate context the phenomenon in question" (Bechtel 2009, 544). The original inspiration for positing mechanisms in the natural world was the analogy with machines or designed artifacts, as the term implies. But when it comes to the paradigm cases of mechanisms studied by science, the entities in question are not artificially designed but "designed" by natural selection. A biological organ, like the heart or kidney, can be thought of as a mechanism, as can a biological cell. If we think of a cell as a mechanism, then it is normally in a state of homeostasis or equilibrium, and moreover, it has a set of regularly instantiated stable properties as a result of remaining in a state of equilibrium. Cells can grow, reproduce, use

---

[14] At this point, one might ask whether the HPC account may be seen to be compatible with essentialism, specifically by considering the mechanism to be the essence of the kind, while possessing the mechanism is the property that defines the kind. This is the understanding of the HPC account that Boyd advocates in some of his work (e.g. Boyd 1999) and it is also favored by some of his interpreters (e.g. Griffiths 1999). Whether or not one puts an essentialist spin on the HPC account of kinds, we shall see in the rest of this section that the emphasis on mechanisms in this account may be misplaced.

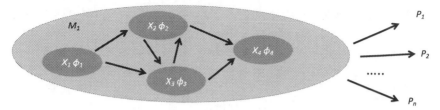

Figure 3 Schematic representation of a mechanism M_1 consisting of entities $(X_1, \ldots, X_4)$ engaging in activities $(\phi_1, \ldots, \phi_4)$ and organized in such a way as to stably and robustly instantiate certain properties $(P_1, \ldots, P_n)$ of the system as a whole (cf. Craver 2007, figure 5.7).

energy from light and glucose molecules, respond to stimuli, and so on. These properties result from the entities that compose them (e.g. nucleus, ribosome, mitochondria) being organized in a certain way and taking part in their respective activities or operations (e.g. storing DNA, translating RNA bases into amino acids, generating energy). Moreover, some of these entities can be considered mechanisms in their own right (e.g. cell nuclei and other organelles), and cells can also be constituents of larger encompassing mechanisms (e.g. hearts and other organs). The entire repertoire of entities and activities that constitute a biological cell is self-regulating and hence remains in equilibrium, thereby ensuring that the properties of the cell are stably and robustly manifested. The basic schema can be handily illustrated in diagrammatic form (see Figure 3).

Biological systems like cells, organs, and indeed organisms would seem to be paradigmatic cases of (nested) homeostatic mechanisms that have stable and robust properties, albeit loosely clustered ones. The properties may only be loosely clustered (polythetic) because many biological mechanisms are not surefire and they are contextually sensitive, manifesting different properties under different conditions. This conception would seem to capture kinds like *cell, cell nucleus, neuron, kidney, heart,* and *organism*. What about other biological kinds, like the species *Drosophila melanogaster* or the genus *Canis*? If individual organisms are thought of as complex systems composed of mechanisms, which are in turn composed of other mechanisms, it is tempting to think that the kinds to which these organisms belong can also be identified by their underlying mechanisms. In the case of the kind *D. melanogaster*, the mechanism, or at least part of the mechanism, might be the type of DNA molecule that is thought to be distinctive of that particular species. Each species would then be distinguished by the genetic code peculiar to that species, which would be causally responsible for generating the phenotypic properties of that species (e.g. morphology, behavior) in every individual member. Unfortunately for this conjecture, biologists do not think that there is a unique genetic sequence distinctive of each biological species. Moreover, as mentioned in Section 2.2, species are commonly thought of as historical entities that persist from the point of speciation to that of extinction. They are often regarded as diachronic groups of organisms that evolve in certain ways, altering both their genotypes and phenotypes over time. Their intrinsic properties do not remain fixed and are not always held in equilibrium, despite the fact that the process of interbreeding within a species tends to stabilize genotypes and phenotypes to some extent. What if the notion of a mechanism is interpreted more broadly, so that the mechanisms that keep the properties of members of a species relatively stable are natural selection and interbreeding rather than the genetic code in the DNA and associated biological structures? As noted by some critics, this still attaches too much weight to synchronic similarity among members of a biological species and does not take into account polymorphisms within some species (e.g. worker and queen ants), which means that the properties of individual members of a single species can vary greatly and without limit (Ereshefsky and Matthen 2005). Defenders of the HPC account have responded to these criticisms by emphasizing the importance of underlying mechanisms, external constraints, and basic causal properties as opposed to surface properties when it comes to individuating natural kinds (Wilson, Barker, and Brigandt 2007, 211). Nevertheless, the emphasis on mechanisms that keep properties in equilibrium may be too stringent and may not fit many focal instances of natural kinds.

The HPC account stresses the importance of mechanisms to natural kinds, but this emphasis is misleading in certain ways. Mechanisms may be prevalent in many kinds in the biological domain as well as in the realms of engineering and technology, but they do not seem as relevant in basic physics and chemistry. Some critics of the HPC account have pointed out that many paradigmatic natural kinds, like kinds of elementary particles, cannot be considered to consist of clusters of properties held in equilibrium by a mechanism (Chakravartty 2007). As far as we know, there is no mechanism inside an electron that gives rise to its properties, as the inner workings of a clock account for its timekeeping properties. It would also be a stretch to think of elemental atoms as being or having mechanisms, at least on most current characterizations of mechanism (see Glennan 2017 for a very broad understanding of the concept, but see Ross 2020 for an opposing view). An atom of beryllium is not exactly an entity that is organized in such a way as to produce regular changes, or an organized set of parts that perform different operations, to paraphrase the two characterizations of mechanism quoted earlier.

Even in the biological world, the HPC account has limited applicability, at least if we insist on individuating kinds in terms of their underlying mechanisms. It is often the case that diverse mechanisms generate the same suite of properties, resulting in what appears to be a unitary kind supported by different kinds of mechanisms. For example, if we consider that *female* and *male* are kinds in the animal kingdom, this does not sit well with the HPC account, since we know that the genetic mechanisms that lead to sex differentiation are quite diverse in different species and lineages. In humans, sex differentiation is controlled mainly by the X and Y chromosomes, but in many reptile species, sex differentiation is under environmental control (often by temperature), and in other lineages there are other genetic, chromosomal, and environmental causes for sex differentiation. In fact, the kinds *female* and *male* may be good examples of *functional kinds*, members of which may not share an underlying mechanism (cf. Khalidi 2020; and see Section 3.2). Rather, they participate in other types of causal structures. There are certain macroscopic selection pressures that favor the persistence of sexual dimorphism across a very diverse collection of animal species, and it is underwritten by a wide assortment of different mechanisms in these lineages. This situation seems quite common in the biological realm. Natural selection often "conspires" to find different underlying mechanisms for achieving the same overarching functional properties. Moreover, this is not just a prevalent feature of biology. A wide array of other kinds of macrophenomena are also multiply realized by underlying microstructures in diverse sciences from fluid mechanics to geology. Many multiply realized kinds can be considered functional kinds, whereby the same overall causal function is achieved by different structures or mechanisms. As some philosophers have pointed out, this means that functional kinds do not "fall within the scope of HPC theory" (Ereshefsky and Reydon 2015, 974). Ereshefsky and Reydon use the example of kinds of *genes*, which are identified by the roles that they play in producing various molecular expression products, to make their case. Since the same DNA sequence can be involved in producing different products in different contexts and different DNA sequences can produce the same products in different organisms or different species, we (again) have a mismatch between underlying mechanisms and their corresponding kinds. (In Section 3.2 we will take a closer look at functional kinds and compare them with other kinds of kinds.)

These cases convey an important lesson concerning the different types of causal networks that are associated with kinds and they suggest that real kinds can have heterogeneous underlying causal mechanisms. Perhaps for this reason, Boyd (1989, 16) sometimes suggests that homeostasis need not be understood literally and that mechanisms need not generate the properties in the cluster: "Either the presence of some of the properties in [the cluster] *F* tends (under appropriate conditions) to favor the presence of the others, or there are underlying mechanisms or processes which tend to maintain the presence of the properties in [the cluster] *F*, or both."[15] But if we don't insist on the presence of a mechanism keeping the properties in homeostasis, we are left with a very different picture of kinds. In Boyd's final paper on natural kinds, he stressed a foundational idea that seems more fundamental to his account of kinds than equilibrium-inducing mechanisms. In that paper, he made it clear that he didn't take all natural kinds to be HPC kinds, or as he put it, "NK $\neq$ HPC" (Boyd 2021; cf. Magnus 2014). Instead, in some of his work, Boyd emphasized that our inferential practices and linguistic categories must be accommodated to "relevant causal structures" (Boyd 2021, 2871; cf. Boyd 2000), without insisting that these causal structures be tantamount to homeostatic mechanisms. He calls this position "accommodationism" since it revolves around the idea that our concepts and theories (our "inferential architecture," as he dubs it) should aim to accommodate the causal structure of the world. Specifically, he writes that the "naturalness [of kinds] consists in a certain accommodation between the relevant conceptual and classificatory practices and independently existing causal structures" (Boyd 2000, 57).[16] Therefore, in the next section, I will investigate what happens when we widen our focus and consider a diverse set of causal structures to correspond to kinds.

2.4 Simple Causal Theory

If we relax the HPC account to allow for a variety of ways in which the properties associated with a kind can be causally linked, we end up with an alternative account of natural (or real) kinds. In some cases, a mechanism is responsible for the properties of the kind and gives rise to them regularly. In other cases, one property regularly causes others, which in turn cause others. In yet other cases, one iteration of a property causes the instantiation of a second property, which then causes another iteration of the first property, and so on. These and other causal structures seem to be found in many instances of the natural kinds that are salient in investigations across a wide array of sciences (see Figure 4). The HPC theory can be thought of as a special case of a broader causal account of kinds.

---

[15] Compare the following exegetical remark on Boyd's views by Ereshefsky and Reydon (2015, 971): "One should not read too much into the term 'mechanism' in 'homeostatic mechanism,' however. Boyd allows ... all sorts of interactions and processes to underwrite kinds. A homeostatic mechanism can be anything that causes (in the broadest sense of the term) a repeated clustering of properties."

[16] But Boyd also stresses that this is not a standard realist position since it accords some role to human interests in the individuation of kinds: "On the accommodationist conception natural kinds are mind-dependent social constructions" (2021, 2871). I will return to this issue in Section 3.4.

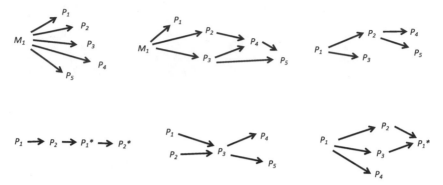

Figure 4 Schematic representations of various possible causal structures associated with natural kinds, only some of which involve mechanisms (cf. Craver (2009, 583), Keil (2003, 370), Ahn et al. (2001, 63)).

The importance of causation to natural kinds has been appreciated from the very earliest writings on kinds by Mill (1843/1882, IV.ii.2) who writes: "The properties, therefore, according to which objects are classified, should, if possible, be those which are causes of many other properties." But Mill was not always clear on this score and some of his earliest critics took him to task for not emphasizing the connection between kinds and causation, since he wrote at times as though the properties that cluster in kinds do so *as a matter of brute fact* (1843/1882, III.xxii.2). Instead, Franklin and Franklin (1888, 84) point out: "The true view of the case seems to us to flow from the general doctrine of Causation."[17] They go on to say:

> In like manner, if all objects which possess the attributes *a* and *b* are found in nature to possess a number of other attributes in common, we cannot believe that this is a mere coincidence; we are forced to conclude either that the given attributes are accompanied by the others in virtue of a general law of causation, or that the objects have a certain community of origin. (Franklin and Franklin 1888, 84)[18]

The connection between causation and kinds has also been flagged by a number of philosophers who have written on kinds, as can be seen from the following passages, drawn from a range of philosophers across the twentieth century:

> [T]he definition of a kind of substances partly depend[s] on the causal laws which substances of the kind are supposed to obey. (Broad 1920, 33)
> To say that one event caused another is to say that the two events are of kinds between which there is invariable succession ... What I wanted to bring out is

---
[17] For illuminating discussions of Mill and his early critics and commentators, see Magnus (2013; 2014).
[18] The mention of "community of origin" brings up the issue of historical or etiological kinds, which will be discussed later in this section and at greater length in Section 3.3.

just the relevance of the notion of kinds, as the needed link between singular and general causal statements. (Quine 1969, 133)

But if we care to imagine languages in which no special causal concepts are represented, then no description of the use of a word in such languages will be able to present it as meaning cause. Nor will it even contain words for natural kinds of stuff. (Anscombe 1971, 93)

The task of science is to expose the causal structure of the world, by delineating the pre-existent natural kinds and uncovering the mechanisms that underlie causal dependencies. (Kitcher 1992, 104)

The ontological ground of induction for such [real] kinds, the reason that the members have many properties in common, is that they have a few fundamental properties and/or causes in common that account with natural necessity for the others. (Millikan 2000, 18)

But the first explicit mention of a "simple causal theory" of natural kinds appears to come from Craver (2009, 579). He observes that it is possible to drop Boyd's requirement of a homeostatic mechanism and "keep the rest as a simple causal theory of natural kinds." As he elaborates: "According to this view, natural kinds are the kinds appearing in generalizations that correctly describe the causal structure of the world regardless of whether a mechanism explains the clustering of properties definitive of the kind" (Craver 2009, 579).

The simple causal theory (SCT) of kinds affirms that kinds can be identified with clusters of properties, but insists that kinds are not *mere* clusters of properties. Causation is what glues these properties together (rather than, say, convention or arbitrary association). Hence, unlike essentialism and like the HPC theory, the simple causal theory of kinds supplies an answer to the question of the "stickiness" of the properties associated with kinds. This view also recognizes that there can be a variety of different ways in which properties can be causally related, leading to a range of different causal structures. The kinds in these cases are properties or sets of properties that are causally correlated with other properties, resulting in causal structures or hierarchies. In these cases, we tend to identify the causally prior properties with the kind itself, as when we say that beryllium is the element with atomic number 3, since that is the property that is causally responsible for the other properties associated with the kind. But this is an oversimplification for several reasons. First, many of the properties associated with beryllium depend also on atomic weight, not just atomic number. Second, many of the properties are only manifested when beryllium atoms are found in large quantities and configured in certain ways (cf. Bursten 2020). A single atom of beryllium cannot be said to have a melting point or density (at least, not the same density as macroscopic samples of beryllium). Third, the typical properties associated with beryllium generally require the presence of certain background conditions to be instantiated. For example, beryllium has high rigidity, but only across certain temperature ranges. For these and other reasons, identifying each kind of chemical element solely with an atomic number, or each kind of chemical compound with a chemical formula (e.g. water is H_2O), distorts a more complex reality. But with that caveat in mind, kinds can be thought of as networks of causally related properties arranged according to causal priority.

Perhaps a better way of elaborating the simple causal theory of kinds is to think of kinds as being parts of recurring causal networks, or to coin a phrase, "nodes in causal networks" (NCN; cf. Khalidi 2018). Here, the nodes are causal properties that are linked to many other properties and tend to initiate causal chains. Since causal relationships can be represented by causal graphs, natural kinds can be described as highly connected vertices in directed causal graphs. The causal graphs often also contain properties associated with background conditions, which together with the "primary" or "core" properties of natural kinds result in the "secondary" or "derivative" properties. This theory, which can be considered to be one way of articulating the SCT, and an elaboration of it, also provides a link between the nature of the kinds themselves and the features of the categories, since causation is the underlying basis for projectibility.

A kind is a recurring causal network rather than a unique causal structure that is only instantiated once in the history of the universe, like the Andromeda Galaxy or the French Revolution. Of course, some kinds may be instantiated many times, such as hydrogen atoms and galaxies, while others may only occur relatively few times, such as roentgenium atoms or revolutions (assuming that galaxies and revolutions are real kinds). Especially outside the domain of fundamental physics, causal networks are not always replicated precisely, and the clusters of causal properties may be instantiated somewhat differently on different occasions. If a causal network is interpreted as a cluster of properties arranged in such a way as to exhibit the causal relations between them, how much of that network needs to be manifested on each occasion for a certain kind to be instantiated? There does not seem to be a clear answer to this question in many cases, particularly in the special sciences. Vagueness is associated with the boundaries between many natural kinds, and this means that some individuals (objects, events, processes, and so on) may be intermediate between kinds. But that seems to be a feature of the universe that any theory of kinds will have to grapple with. There are fuzzy kinds in nature, and any naturalistic theory of real kinds must take that into account.

There are several objections that might be raised against the SCT (or NCN) account of kinds, which will help elaborate the theory and make it more substantive. First, it should be noted that the SCT does not fully account for the most fundamental kinds in the universe, at least on our current best theory: quarks and leptons. These kinds of particles are characterized by their basic properties, mass, charge, and spin, but there is no causal account of why these specific properties co-occur or stick together. If these particles are truly fundamental, then the coinstantiation of their most basic properties would seem to be a matter of brute fact rather than causality. Of course, it is still the case that the properties of each kind of elementary particle, either singly or in combination, lead to the instantiation of the other properties associated with that kind of elementary particle. For example, the fundamental properties of electrons account causally for why they attract protons, deflect in an electric field, are annihilated when they encounter positrons, and so on. Still, there is no causal story that explains the coinstantiation of the most basic properties of electrons (mass, charge, and spin), and we have no reason to think that there will be one, at least according to our current understanding of elementary particle physics. This also applies to the other fundamental particles, including both quarks and leptons. Proponents of SCT might make a couple of points in response. First, even though there is no causal account of the fundamental properties of elementary particles, those properties are causally linked to their other properties; indeed, they cause all their other properties, either directly or indirectly, as already mentioned. This means that causation gives at least a partial account of the totality of properties associated with each kind of elementary particle. Another reply is that even though we currently have no causal story for why the properties of elementary particles are associated, that does not mean that there isn't one. In fact, we might distinguish two possibilities. On the first, causation would "bottom out" at the level of quarks and leptons (or at some more fundamental level). In this case, it would be hardly surprising that we're unable to give a causal account of the most fundamental kinds in the universe, since causation comes to an end somewhere. The other possibility is that causation goes "all the way down," in which case there might be a causal account of the clustering of properties at all levels ad infinitum, and SCT would apply to those kinds, after all.[19]

A second objection harks back to the quotation from Franklin and Franklin near the beginning of this section. Recall that they criticized Mill for not being forthright enough about the fact that the properties associated with kinds are causally linked. But they also noted that when entities that share certain properties, say P_1 and P_2, also share a number of others, then this may be due

---

[19] The possibility of "infinite descent" is taken seriously by some physicists and is also discussed from a philosophical perspective in Schaffer (2003).

either to causation or to "a certain community of origin." This second disjunct brings up what some philosophers consider to be a different underlying basis for kinds. It can be argued that members of biological species and other kinds share properties not because those properties are causally linked to each other, at least not straightforwardly. The giraffe's neck and camouflage do not seem to be directly causally linked. They may be thought to be effects of a common cause, namely the DNA characteristic of their species, but this just raises the question: why is it that giraffes have similar DNA (bearing in mind the caveats about interspecies genetic diversity from the previous two sections)? The obvious answer is that it is a result of the fact that they are ultimately descended from common ancestors. Beryllium atoms do not all have the very same origin (apart, of course, from the Big Bang, which is the common origin of all kinds), whereas giraffes and fruit flies do. This suggests that for at least some kinds, sharing properties is a result of a historical process or "community of origin" (as Franklin and Franklin put it), rather than (or in addition to) synchronic causes. Is this, then, an independent reason for property sharing among individuals belonging to the same kind? Although there is an important subcategory of kinds, historical or etiological kinds, that are individuated at least in part by their history or origin, proponents of the causal theory of kinds could still point out that the history or origin involved is a *causal* one (cf. Khalidi 2021). The lineage of a biological species represents a causal process or trajectory that unfolds over evolutionary time, and it yields individuals with similar properties because it is a copying process (Millikan 1999, 54–55). Millikan has distinguished such "historical kinds" (or "copied kinds," cf. Elder 2004) from other kinds of kinds, but the fact remains that they also involve causal pathways that result in shared properties. (We'll return to the issue of etiological kinds in Section 3.3.)

Some philosophers might also object to a simple causal theory of kinds by saying that it is too permissive or pluralist (recall Section 1.3). By allowing kinds to correspond to properties that are causally linked to other properties, we may risk admitting too many kinds into our ontology. As Pöyhönen (2016, 150) asks rhetorically, "if all causally sustained groupings qualify as natural kinds, does this not lead to an explosion in the number of natural kinds?" To address this issue, it may be worth harking back to some of the earliest writers on kinds. Mill's view was that kinds should be characterized by an "indefinite" or "inexhaustible" number of properties. By contrast, C. S. Peirce (1901) objected to Mill on this point and seems to have been quite content to allow kinds to be associated with just two coinstantiated properties. Peirce (1901) defines a "real kind" as follows: "Any class which, in addition to its defining character, has another that is of permanent interest and is common and peculiar to its members, is destined to be conserved in that ultimate conception of the universe at which we aim, and is accordingly to be called 'real.'" Mill's insistence on inexhaustibility seems both unwarranted and unrealistic (see Khalidi 2013, 112ff), but does that mean that we should allow any property that regularly and reliably causes one other property to correspond to a kind, as Peirce appears to do? Is that too low a bar and would it open the floodgates to dubious candidates? An advocate of the SCT might allow such kinds as long as the properties are causally linked. One property that regularly causes another can be considered a kind on this theory since it represents a type of very simple causal structure, albeit one that does not appear to be very common in our universe. In more typical cases, one or a cluster of properties causes a whole slew of others, so we are not likely to admit too many two-property kinds into our ontology. This suggests that fears of an explosion may be overblown, particularly since, as argued in Section 1.3, pluralism about kinds is certainly compatible with realism.

A fourth objection pushes in the opposite direction. Some critics of the SCT (or NCN) approach to kinds have protested that it is too restrictive in not admitting many bona fide scientific kinds. The claim is that there are many kinds that are theorized about in science and play a role in our inductive and explanatory practices yet are not causally based. These kinds are left out by an account that privileges causation and causal links between properties. This criticism has been made forcefully by Ereshefsky and Reydon (2015, 970), who write that philosophy of science "needs an account of kinds that better captures the variety of classificatory practices found in science." Though their criticism is aimed primarily at the HPC account of kinds, they also think that insisting on causal linkages between the properties associated with kinds is too restrictive. Their primary counterexample comes from microbiology, specifically the Phylo-Phenetic Species Concept (PPSC), which is used to classify bacteria. They observe that the "aim of the PPSC is to capture stable kinds that have clear identity conditions" and that it uses a number of empirical parameters to do so, but not causal mechanisms (Ereshefsky and Reydon 2015, 973). However, this purported counterexample to the SCT has been challenged by Lemeire (2021, 2917), who argues that this classification scheme is merely practically convenient and that even scientists who use it consider it a "pragmatic species concept" that is both "arbitrary" and "anthropocentric." As pointed out in Section 2.1, if some taxonomic schemes in the sciences are adopted out of convenience or expediency, they may not be taken as guides to the real kinds. Similarly, Santana (2019) demonstrates that the main classification system in mineralogy does not divide the mineralogical domain into natural kinds. While admitting that some mineralogical species may correspond to natural kinds, he argues that the main taxonomic system in mineralogy is anthropocentric and geared to practical needs and hence does not aim to uncover real divisions in nature. In such cases, it seems that the right thing to say is not that the SCT is too restrictive but that some scientific taxonomic schemes do not aim to reveal real kinds.

Before concluding the discussion of the SCT, it's worth comparing it briefly with the essentialist and HPC accounts, at least when it comes to their relation to realism. An essentialist account can be understood as supremely realist or perhaps hyperrealist. It posits real essences as objective metaphysical structures that underwrite natural kinds. Meanwhile, Boyd considers the HPC account, as well as the accommodationist conception that it belongs to, to be "construction-ist" or "mind-dependent," and he explains that "on the accommodationist conception ..., natural kinds and their definitions are discipline-or-practice relative and are thus not 'mind independent'" (Boyd 2021, 2889). The SCT seems to represent a position intermediate between the essentialist and HPC accounts, since it doesn't posit metaphysical essences, but it takes natural kinds to be objective features of reality as the products of causation. Now it may be that causation itself is not fully objective, and if so, then on the SCT, kinds would have to follow suit. But if we assume a realist account of causality, then the SCT would be a realist account of kinds. The SCT is also compatible with some aspects of the accommodationist conception outlined by Boyd, at least if one interprets it more broadly in terms of engineering our categories to accom-modate aspects of reality, specifically the causal structure of reality. Given the advantages that the SCT seems to offer over essentialism and the HPC theory, I will be assuming it in what follows, in looking at different kinds of kinds and specific examples of kinds in various domains.

The theories of kinds presented in this section all propose some kind of metaphysical basis for real kinds (essences, homeostatic mechanisms and property clusters, causal structures or networks). But some philosophers have advocated steering clear of metaphysics when it comes to kinds. An "epistem-ology-only" theory of kinds would suspend judgment on the underlying meta-physics of kinds, focusing instead on explanatory categories in the sciences. Strictly speaking, such a theory would bracket *kinds* in favor of *categories*. This is the type of view put forward both explicitly and implicitly by several authors, including Magnus (2012), Slater (2015), Franklin-Hall (2015), and Ereshefsky and Reydon (2015). On one such view, kinds correspond to *stable property clusters* (SPC; see Slater 2015) without a commitment to providing a metaphysical undergirding for the stability of the cluster. But there are costs to remaining agnostic about the metaphysics of kinds, since some of the epistemic features of kind categories, such as explanatory value, are grounded in the metaphysics, such as relations of causal priority, and cannot be fully exploited without an account of the underlying metaphysical picture. Lemeire (2021) argues that while mere clustering may be enough for inductive inference, it's not enough for explanation or categorization (the practice of assigning kind membership to individuals).

3 Kinds of Kinds

The account of natural kinds sketched in Section 2.4 is realist, at least assuming that causation is a real relation, but pluralist in that it allows kinds to be grounded in a variety of different causal structures in the world. It is less pluralist than some accounts of kinds, since it requires kinds to be rooted in causality but does not require essences and does not invoke modal necessity. It does not even require homeostatic mechanisms. Despite the fact that it is a unitary account, which grounds kinds in causation, the simple causal theory also allows for a variety of causal structures to give rise to kinds. In this section I will be exploring various different ways in which causal configurations give rise to kinds and the complex systems of kinds that result. In Section 3.1 I will clarify and reject the claim that kinds are arranged hierarchically, supporting the view that real kinds can cut across one another and reinforcing pluralism about kinds. Then, in Section 3.2 I will defend the claim that kinds can be functional or relational in nature and that their causal properties can be multiply realized by their microstructural properties. In Section 3.3 I will explicate another important type of kind that occurs in a number of sciences, namely historical or etiological kinds, which are individuated according to their origins or the histories of their members. Finally, in Section 3.4 I will try to show that real kinds can be mind-dependent in various ways, which supports the idea that there can be real kinds in the psychological and social sciences.

3.1 Crosscutting Kinds

In Section 1.2 we encountered the idea that the kinds investigated by the sciences are discipline-specific or interest-relative. Indeed, it could be said that kinds are sometimes even relative to specific *sub*disciplines or *research programs*. The kinds invoked by planetary astronomy are different from those deployed in historical linguistics; they may even be disjoint. But I also argued there that the fact that kinds pertain to specific disciplines does not imply that they are conventional or arbitrary in nature. As long as each discipline is interested in identifying the real kinds in its domain, those kinds will be objective features of reality. In light of the discussion in Section 2.4 and assuming something like the simple causal theory of kinds, if a specific research program is successful in identifying real kinds, those kinds will be aspects of the causal structure of the world, even though they are only investigated by that specific research program. If we think of disciplines, subdisciplines, or even research programs as trying to identify certain causal networks that are somewhat self-contained or can be studied in relative isolation, then the causal nodes in these networks are in some sense proprietary to the domains investigated but also fully objective. In order to do science, we have to zero in on some partial aspect of a "causal thicket" (Wimsatt 2007), but the fact that a kind category features only in one corner of science doesn't mean that it is merely relative or unreal. There is no conflict between the claim that the kinds that are correctly identified by each research program are distinctive of that particular program and the claim that these kinds are real. True enough, *photosynthesis* is not studied in psychology, and *unemployment* is not a construct of microbiology. But that doesn't mean that these categories pick out real kinds only in relation to botany and economics, respectively. Insofar as they are real kinds, they are real kinds full stop; it's just that they are not the focus of all sciences.

Of course, it's also true that a single kind can figure in quite different disciplines, for example *ionization* in elemental chemistry and in neuroscience. In some of these cases, the kinds feature in orthogonal or somewhat askew causal processes. In other cases, different disciplines or subdisciplines can study more or less the same domain and posit somewhat different kinds because they have different explanatory or epistemic interests. Psychology and neuroscience are both interested in human behavior and study individual human beings as perceivers, thinkers, and decision-makers. But while psychology posits such kinds as *concept*, *emotion*, and *episodic memory*, neuroscience deals with such kinds as the *hippocampus*, *action potential*, and *long-term potentiation*. Given that they sometimes have different explanatory interests, we might expect that their kinds might not always coincide (cf. Khalidi 2017).[20] Indeed, in such cases, different disciplines or subdisciplines can identify systems of crosscutting kinds in the following precise sense. Categories or kinds crosscut one another when they overlap and neither is wholly contained within the other. To take a simple example, consider two different biological subdisciplines studying the animal kingdom, like phylogenetic zoology and developmental biology. The first might divide animals into different species (e.g. *honeybee*, *American lobster*), while the second might (also) split them into different developmental morphs (e.g. *larva*, *adult*). If we take a particular honeybee larva, *a*, and a particular American lobster larva, *b*, both of them belong to

---

[20] This is not to deny that there can be disciplines or subdisciplines, like cognitive neuroscience, that supply bridges or links among such disciplinary taxonomies, but crosscutting often persists despite these links. For more on such "interfield theories" and an argument that they don't always supply reductions, see Darden and Maull (1977).

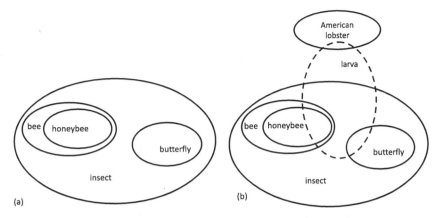

Figure 5 Two Venn diagrams showing (a) a hierarchical, noncrosscutting system of phylogenetic categories, and (b) the same system of categories with a crosscutting category (*larva*) superimposed on it, which includes some but not all members of the original categories as well as some nonmembers.

the kind *larva*, and if we take *b* and an American lobster adult, *c*, both of them belong to the kind *American lobster*, but none of the kinds mentioned comprises all three individuals, *a*, *b*, and *c*. These categories crosscut each other in the sense that they can't be arranged in a nested hierarchy. The developmental categories crosscut standard phylogenetic categories like species, genus, and family (see Figure 5). To take another example of crosscutting categories, consider the categories of chemical substances and the phase categories, *solid*, *liquid*, and *gas*. A sample of liquid water and a sample of ice both belong to the kind *water* (H_2O), and a sample of ice and a sample of solid table salt (NaCl) both belong to the kind *solid*, but none of these kinds include all three samples.[21]

The idea of a neat hierarchy of kinds is an attractive one and it may indeed be traceable back to the notion of the "Great Chain of Being," whose roots lie in ancient Greek philosophy and whose most developed incarnation emerged in medieval Europe. Though it is widely thought to have been discredited in the Enlightenment, this conception of the world may live on in the assumption that natural kinds are arranged hierarchically. Yet, the examples cited earlier, as well as many others, suggest that that is not the case (cf. Khalidi 1993; Khalidi 1998; Tobin 2010; Havstad 2021). But the idea of a nested hierarchy of kinds need not

---

[21] For a recent wide-ranging discussion of crosscutting kinds and an in-depth study of crosscutting kinds in biochemistry, see Havstad (2021). For an early defense and critique of hierarchy among natural kinds, see, respectively, Thomason (1969) and Kahane (1969).

go so quietly, since a question may be raised as to whether crosscutting categories are genuine scientific ones and are even putative candidates for being real kinds. To illustrate using the preceding examples, defenders of hierarchy might deny that crosscutting occurs on the grounds that the developmental categories in biology and the phases of matter in physics and chemistry do not correspond to real kinds. In this case they can be considered "phase sortals," which apply to entities at just some points in time and not others. Hence, they are not genuine kinds. They may conclude that crosscutting can occur among some scientific categories but not among the categories that correspond to kinds.

However, it's not so easy to dismiss the biological developmental categories or the phase categories in physics and chemistry. At least according to the "simple causal theory" of kinds outlined in Section 2.4, these categories would seem to be excellent candidates for kinds. A category such as *larva* is projectible, features in explanations, and is capable of being used to predict and manipulate nature. Moreover, these epistemic features of the category are rooted, as in other cases, in the causal structure of the world, whether synchronic causal processes or etiological (causal-historical) reality. Like many biological phenomena, the life stages of insects, crustaceans, amphibians, and others evolved as a polymorphism by natural selection. It is an adaptive feature of some organisms and confers certain advantages on lineages that have this feature. Also, each developmental stage has distinctive features of its own. Larvae are adapted for feeding (by contrast with more mature organisms that are focused on reproduction) and they are particularly adept at finding food sources. They evolved to have features that endow them with causal powers common to larval morphs across biological lineages. As for the phase categories, they are, of course, ubiquitous in numerous sciences that are concerned with whether matter is in a solid, liquid, or gaseous state. Many generalizations can be made about diverse substances in these respective phases, which are based on the causal properties associated with each of these phases, for example, the relative distances between the atoms or molecules and the amount of kinetic energy they have. Without going into more detail, it is safe to say that the phases of matter are not just categories that pertain to folk taxonomies, but they play an important role in a range of natural sciences. Therefore, in both cases, these are robust systems of scientific categories that are grounded in causal structures in reality and cannot be ignored by friends of the hierarchy thesis. Generally speaking, science is replete with categories that crosscut each other and are also prime candidates for corresponding to real kinds, and this casts serious doubt on the idea that kinds are arranged in a hierarchy. It is also worth emphasizing that crosscutting categories do not always correspond to phase sortals. Havstad

(2021) discusses the case of nuclear receptor proteins, which act as transcriptional activation switches inside the nuclei of cells, arguing that they can be classified in at least three ways: procedurally typological, phylogenetic, and effect-based typological. She concludes that "there is more than one taxonomy of natural kinds" and that these taxonomies crosscut one another (Havstad 2021, 7691).

If real kinds are not arranged in a neat hierarchy and they can instead crosscut one another, what implication does this have for the nature of kinds? For one thing, it provides yet another source of pluralism in addition to those explored in Section 1.3. But the denial of hierarchy can also be seen to be implicated in an influential attempt to derive essentialism (see Khalidi 1993). The reason is that some derivations of essentialism require an assumption to the effect that each individual entity in the universe belongs to one unique natural kind, or if not, to a hierarchy of noncrosscutting natural kinds. Unless one can single out an individual and use it to anchor a kind (or a set of hierarchical kinds), then one cannot use that individual to fix the essence of a kind across possible worlds (cf. Crane 2021).

Finally, the phenomenon of crosscutting may contain a lesson when it comes to the age-old question of lumping and splitting in science. Should scientists aim, whenever possible, at consolidating categories and grouping things together in more comprehensive classes that bring out similarities among seemingly diverse entities (lumping)? Or should they instead seek to divide entities into subcategories, each of which is distinguished by greater numbers of similarities (splitting)? If crosscutting is the norm and pluralism is widely accepted, a choice between lumping and splitting may not need to be made. At least in some cases, we can lump for some purposes and split for others, depending on our interests. We can emphasize certain divisions in some of our inquiries and ignore them in others, without having to forgo one classification for the sake of the other. Strictly speaking, lumping and splitting apply within a hierarchical taxonomic system, as when one theorist focuses on species and another on genera, but it can also be seen to pertain to crosscutting systems, since some taxonomic schemes might group entities into smaller classes than others.

3.2 Functional Kinds

A contrast is often drawn between kinds that are characterized by intrinsic properties and those distinguished by extrinsic or relational properties. Many relational properties and kinds are traditionally not thought to be natural or real. It is not a simple matter to distinguish intrinsic from extrinsic (or relational) properties, but we can assume a rough-and-ready understanding of relational properties for these purposes.[22] Toy examples of relational classes like the class of *all and only objects lying within one kilometer radius of my nose* or *all and only things that have been in my car* are sometimes cited as arbitrary categories that should not be expected to correspond to kinds. But that need not be the case for all relational categories, and there are arguably many scientific categories that are relationally based but not so easy to dismiss. Relational properties are causally efficacious when the relations in question are embedded in a causal system, participate in regular causal processes, and when the input-output relations associated with those properties are generally uniform. This uniformity in causal-functional profile – rather than their intrinsic properties or features – is what grounds their being kinds. Thus, an important subclass of relational kinds is often dubbed "functional kinds" (Weiskopf 2011a; 2011b). We have encountered such kinds before in the context of discussing the HPC theory of kinds (Section 2.3), specifically in considering the idea that some biological kinds are such that diverse causal mechanisms undergird uniform causal-functional properties, as in the case of the putative biological kinds, *female* and *male* (cf. Khalidi 2020). In these cases, it was claimed that the strict version of the HPC theory has a hard time accounting for their kindhood, given that they are not individuated by a single type of mechanism that holds their associated properties in equilibrium. Rather, in these cases as well as many others in the biological domain, natural selection may recruit a variety of mechanisms to perform the same or highly similar functions. The causal properties associated with these kinds are not a result of uniform intrinsic mechanisms, but rather they are a matter of how they function in a causal "economy."

Another example might help make the point, this time from the realm of artifacts, which may be the archetypal functional kinds. Consider thermostats, all of which function to ensure that the temperature in a given space (e.g. room, house, airplane cabin) remains at a certain determined value. When the temperature drops below that value, the device switches the heater on, and when the temperature reaches the requisite value, the device switches the heater off again. Notwithstanding this uniformity of function, thermostats come in all shapes and sizes, from those based on bimetallic strips to those relying on thermocouples, and a host of others in between. Artifactual devices like thermostats may be among the clearest cases of functional kinds that are multiply realized with different types of components, structures, and materials. Moreover, their causal

---

[22] Some of the complications involved in characterizing intrinsic and extrinsic properties are discussed in Lewis (1983) and Langton and Lewis (1998).

powers are a relational matter since they act to switch the heater on or off depending on the ambient temperature and provided they are hooked up in the right way to the heater's controls.

But this pattern obtains not just for biological and artifactual kinds, since functional kinds can also be drawn from physics and chemistry (as well as from the cognitive and social sciences, among other disciplines). In fluid mechanics, a *Newtonian fluid* is a kind of fluid whose viscosity remains constant for a given temperature and pressure and is independent of the magnitude of the force acting upon the fluid. This kind includes a wide variety of substances with various underlying molecular structures and chemical bonds. The kind is not characterized in terms of a certain microstructure and it is, moreover, relationally characterized in terms of its behavior in response to an applied force. But it is nevertheless a kind with a specific macrocausal profile and there are significant empirical generalizations involving Newtonian fluids, all of which depend on the causal properties of Newtonian fluids. In chemistry, the kind *polymer* also has some of the basic features of functional kinds, though it is not unequivocally either a functional or a microstructural kind. Polymers are chemical substances with highly variable compositions, including both organic and inorganic compounds, consisting of long chains of molecules, typically in repetitive patterns. This chief property gives them a number of distinct causal properties, which can be used to make generalizations about them and explain certain common chemical behaviors.[23] In these and other cases of functional kinds, there is relative independence of the overall properties of the system from their underlying intrinsic properties, so they are often *multiply realized*, or at least *multiply realizable* (Weiskopf 2011a). The idea is that many different substrates can give rise to or realize the same or similar global functions.

Various philosophers have wondered how it is that diverse microproperties can result in fairly uniform macroproperties, and hence how functional kinds are even possible. When viewed in a certain way, it can seem nothing short of a miracle that dissimilar microproperties can realize the very same macroproperty (cf. Kim 1992). This is an enduring puzzle about the universe that I can't hope to fully address here, but to dispel some of the mystery, it might be useful to consider a common objection in this vicinity. Some philosophers balk at the idea that such multiply realized functional kinds can just be a brute feature of the universe, but they nevertheless allow that design, whether natural or artificial, can give rise to functional kinds (see e.g. Papineau 2010). Thus, the thermostat that relies on the bimetallic strip and the one that deploys the thermocouple can both be instantiations of the same functional kind, *thermostat*, but that is only

---

[23] These and other functional kinds are discussed at greater length in Khalidi (2013).

because humans designed these artifacts to have the same function despite their disparate underlying mechanisms or structures. Similarly, in the biological domain, natural selection sometimes opportunistically favors different mechanisms to perform the same survival-related task, depending on the materials or substrata at hand. For example, two different kinds of fish, Arctic cod and Antarctic notothenioid fish, both have the capacity to produce antifreeze glycoproteins to cope with their frigid environments (Chen et al. 1997). But these antifreeze compounds are manufactured by different kinds of genes in the two biological lineages. Thus, some would allow that the same functional kind can emerge from different kinds of underlying causal structures in cases of natural or artificial design, but not in general. However, the problem with this half-hearted vindication of functional kinds is that it ignores the fact that the only reason that natural selection or human engineering can exploit these resources is that they are already there for the taking in our universe. Biological and human design do not create the conditions that enable the evolution or engineering of these kinds; they merely tap into existing properties. To revert to an earlier example, if Mother Nature or a human engineer needs to locate a Newtonian fluid with a certain viscosity to solve a certain design problem, they may succeed in finding two different microstructural ways to address it, with different microproperties undergirding the same macroproperty or function. But these functional kinds are present in the universe regardless of natural selection or artificial design.[24]

3.3 Etiological Kinds

In Section 2.4, Franklin and Franklin, two early critics of Mill's account of kinds, were quoted as putting forward two candidates for the causal basis of kinds. They held that whenever a number of individuals possess a set of attributes in common, this is either because there is a "law of causation" linking those attributes, or else the attributes have a common origin (Franklin and Franklin 1888). The first case is the one that we have focused on so far. In most of the cases that we've discussed, there is a causal link between some properties and others, and this is what gives rise to a causal cluster of properties. This causal structure has been represented diagrammatically in terms of nodes

---

[24] Here, I am eliding an important distinction between two kinds of functional kinds. The first have their functions by virtue of playing a certain causal role in relation to other kinds, while the second have their functions by virtue of having been selected to play a causal role by natural or artificial selection (these are often dubbed proper functions). But the latter depend on the former, since "the proper functions of a biological trait are the functions it is assigned in a [causal-role] functional explanation of the fitness of the ancestral bearers of that trait" (Griffiths 1993, 410).

in a network that are joined to other nodes in that same network. But these causal links can also be summarized in terms of causal laws or generalizations[25] that relate the properties in the cluster to one another. The diagrammatic representation is preferable to a propositional or sentential one, since it brings out the causal structure more clearly and contains more information than a simple causal generalization. But it is now time to look more closely at Franklin and Franklin's second reason for causal clustering, namely commonality of origin. It's important to get a better understanding of this other basis for kinds and to determine whether it is a genuinely different underlying basis for kinds.

Some of the paradigm cases of real kinds would seem to have certain properties in common because they all descend from some common origin or have the same history. Members of a biological species typically share many common properties. Almost all members of the species *Drosophila melanogaster* have six legs, two wings, red eyes, and black stripes across their abdomens, can sense air currents with hairs on their backs, and feed on fermenting fruit, in addition to numerous other anatomical, physiological, and behavioral traits. It is tempting to see this cluster of properties as being due to a common internal mechanism, which would include the distinctive genetic sequence associated with the species *D. melanogaster*. This would make a biological species an instance of the HPC theory of kinds (which is itself a special case of the SCT). In this case, an intrinsic causal mechanism, incorporating the genetic sequence contained in DNA code, gives rise to a suite of synchronic causal properties, including the ones just mentioned. But, as argued earlier in the discussion of the HPC theory of kinds (Section 2.3), this is a misleading way of thinking about species as kinds. For one thing, there is no unique genetic sequence distinctive of each biological species. More importantly, members of a species do not all have the same genotypes and phenotypes, and these change over time under certain selection pressures. Hence, rather than identifying biological species with clusters of synchronic causal properties generated by a common mechanism, they can be thought of as branches in a phylogenetic tree. Each species is a lineage that begins at the point of speciation (which may be somewhat vague) and ends at the point of extinction. That means that species can be thought of as historical entities with a common origin. Members of a species tend to share properties (albeit imperfectly) because of that common origin and a history of descent. If this is a better picture of biological species kinds, then it seems to

---

[25] I would prefer the terminology of causal generalizations to that of causal laws, since the latter suggests exceptionless universal generalizations, and most causal links in science are riddled with exceptions. For a more pragmatic approach to scientific laws, see Mitchell (2000).

constitute a different basis for something being a real kind, namely a diachronic or historical basis rather than a synchronic one.

Millikan (1999; 2005) has perhaps done most to promote the idea that there are "historical kinds" and that they are importantly different from "eternal kinds," which are the more usual real kinds, like chemical elements or compounds (see also Griffiths 1994). Millikan (1999, 54) writes: "The members of these [historical] kinds are like one another because of certain historical relations they bear to one another (that is the essence) rather than by having an eternal essence in common." Millikan (2005, 307–308) also identifies three causal factors that lead members of a historical kind to be similar to one another (broadly speaking). First, there is a process of reproduction or copying, with members of the kind having been produced from one another or from the same models. Second, members of the kind are produced in response to the same environment. Third, some function is served by members of the kind, where "function" is understood in the biological sense as an effect raising the probability that its cause will be reproduced. She also thinks that the third causal factor tends to support the first one. That is, individual members are copied precisely because they serve a function. Moreover, these three causal factors are typically combined. This basic blueprint seems to fit biological species, in addition to other historical biological kinds like homologies, but it can also be applied to social kinds like artifacts (e.g. *car*, *screwdriver*) or institutions (e.g. *parliament, jury*). In all these cases, members of the kind tend to share properties because they are copied, shaped by the environment, and play a certain role or function. It is the historical process that is fundamental to their identity; the shared synchronic properties are just a consequence of that common historical process (cf. Godman 2021).[26]

Notwithstanding the significant differences between "eternal" and "historical" kinds that have just been emphasized, it is important to stress that historical kinds also correspond to a *causal trajectory*. By identifying them, scientists aim at capturing an aspect of the causal structure of reality, albeit a diachronic rather than a synchronic one. To put it differently, they can also be seen to conform to Boyd's "accommodationist" conception of kinds, mentioned in Section 2.3, according to which our inferential practices and linguistic categories ought to accommodate "relevant causal structures" in the world (Boyd 2021, 2871). Since historical kinds are individuated on the basis of causal trajectories, they can also be understood in terms of the simple causal theory of real kinds

---

[26] See Reydon (2006) for a thorough investigation of the ways in which historical evolutionary processes give rise to individual members of biological species with shared synchronic properties about which one can generalize. His discussion problematizes the simple picture according to which a shared history results in shared synchronic properties.

outlined in Section 2.4. This is why it would be more accurate to refer to such kinds as "etiological" rather than "historical," since they correspond to an aspect of the *causal* history of the universe.

The discussion so far has focused on etiological kinds like biological species, which have two features that are not common to all such kinds. The first is that they can be considered "copied kinds," since they are copied from one another or reproduced (perhaps imperfectly) from a common template. The second is implied by the first, namely that they have the very same origin or trajectory, that is, they have the same *token* history rather than the same *type* of history. To see that there are etiological kinds that do not have the same token history, consider the geological kinds: *sedimentary, igneous,* and *metamorphic rocks.* Sedimentary rocks are classified as such not on the basis of their composition or structure but because they have the same history, though not the very same *token* history. Rather, all sedimentary rocks have the same *type* of history: they are formed when particles are transported by water, wind, or gravity, deposited in one location, and later compacted to form larger rocks. Since they do not have the same token history, they are therefore not copied from one another or from a common template. Type-etiological kinds are classified as such because they have the same type of history, which is not unique but may be repeated at different times and places in the universe.

There can also be kinds that have the same token history but are not straightforwardly copied. Consider a cosmological kind like *cosmic microwave background radiation*, which is a type of electromagnetic radiation all of whose instances originate in an early stage of the formation of the universe. This kind comprises photons with radiation of a certain frequency that have been traveling through the universe since shortly after the time of the Big Bang, and whose existence provides some of the most conclusive evidence for the Big Bang. But there is no sense in which these photons are copies of each other or of some original photon. When it comes to other token-etiological kinds, even when there is some copying taking place, it's not always a straightforward case of some individual a_1 being copied to produce a_2, which in turn is copied to give rise to a_3, and so on. Take something like sexual dimorphism within and across animal species. Females and males can be considered token etiological kinds, since sexual dimorphism seems to have evolved only once in the history of life on earth, but it is not the case that females are copied from other females, or males from other males. Similarly, with other biological polymorphisms, such as the castes found in certain species of insects, for example, *queen, worker,* and *drone ants.* Thus, Millikan's "historical kinds" are a special case of etiological kinds, since they are copied kinds (which also means that they are token-etiological kinds).

At this point, a question could be raised about the point of individuating some kinds historically or etiologically. In the case of copied kinds, as mentioned earlier, Millikan thinks that the historical origin and trajectory account for the synchronic properties of these kinds, so it stands to reason that we would individuate them historically. When it comes to biological species, for example, a history of selection pressures can be cited to explain many of the synchronic features of individual organisms that belong to that species. However, this might raise a question as to the need for historical individuation in the first place. Why not just classify members of such kinds on the basis of their synchronic properties? In these cases, since the history is more fundamental than the shared properties, as already noted, that would seem to provide some grounds for privileging history and historical individuation. Moreover, history can also account for nonshared properties and for variations among members of a kind, since common historical properties can help account for these variations. This may be particularly evident in cases of homologous phenotypic features, such as mammalian forelimbs. The front legs of a cat, wings of a bat, and fins of a porpoise all belong to the kind *mammalian forelimb*, but each exhibits different properties as a result of differences in their specific historical trajectories, despite their shared origin. The common origin and different trajectories can together account for their differing synchronic features, which are adapted to different forms of locomotion, despite some structural similarities.

What about etiological kinds that are not copied or are not even the result of the same token history? Are there good reasons for historical individuation in those cases? There are a couple of points that can be made here. First, even when it comes to type-etiological kinds, the same type of history can be cited to explain at least some of the synchronic properties of the individual members of the kind. Sedimentary rocks can differ radically in their compositions and structures, yet many of them do exhibit layers of sediment that bear witness to their origins. Second, historical individuation of kinds can serve to explain their differing properties when they differ. In these cases, path dependency can help to account for the fact that some sedimentary rocks are sandstone while others are limestone. Even though historical properties may appear causally inert, they reveal causal structure in important ways, and they generally aid in the project of delineating the causal structure of the universe.[27]

A final concern when it comes to etiological kinds, particularly token-etiological kinds, has to do with whether they should be considered kinds at all. Token-etiological kinds are spatiotemporally bounded unique processes in

---

[27] For more on how historical properties and kinds can illuminate causal relationships and generalizations, see Page (2021).

the history of the universe. That means that it is possible to think of them as individuals rather than kinds. This case has been made most prominently for biological species, which some biologists and philosophers of biology have proposed should be considered individuals, not kinds (Ghiselin 1974; Hull 1978). Rather than kinds with members, they can be thought of as spatially and temporally contiguous individuals with parts. While it is possible to think of species as individuals, it is by no means obligatory and it is at least as natural to regard biological species as kinds with individual members as it is to construe them as individuals with discrete parts. As Van Valen (1976) argues, species can be regarded as individuals for some purposes and classes for others, depending on the processes in which they feature. The same goes for other token historical kinds, like homologous phenotypic features.

3.4 Mind-Dependent Kinds

Among the many kinds that the sciences have identified there are those that are dependent on the human mind in some way. Such kinds are particularly perplexing for many philosophers, since dependence on the mind is often regarded as a mark of what is subjective or not real. Can there be real kinds that are also mind-dependent? It is safe to say that if there are any real kinds in the psychological and social domains, then most, if not all of them will be mind-dependent. Kinds in the psychological sciences like *concept, emotion,* and *episodic memory* are obviously mind-dependent since they pertain directly to the mental domain. Moreover, kinds in the social sciences, like *money, government,* and *marriage,* are also plausibly mind-dependent. It is difficult to see how one could have an institution of *money* unless there was a community of minded individuals who either implicitly or explicitly engaged in certain practices, adopted certain attitudes, and interacted with each other in certain ways. These practices, attitudes, and interactions are all ones involving mental states, such as the state of *valuing* one thing over another or *preferring* one outcome over another.

It is tempting to think that the type of mind-dependence at issue when it comes to psychological and social kinds is not the type that would threaten realism about them. But the mind-dependence when it comes to such kinds is central to their identity. For example, a kind like *money* is not just causally mind-dependent, but constitutively so. It is not just that individuals with minds undertake certain actions that give rise to instances of the kind, but rather the institution itself is in some sense constituted by the actions and attitudes that people take. By accepting certain forms of payment for goods and services and not others, deeming some tokens to be counterfeit but not others, and generally conducting themselves in certain ways, human beings bring it about that there is such a thing as money in their community. If members of that community suddenly ceased to accept tokens of money as payment and decided to embrace the barter system instead, money would thereby cease to exist (though its physical manifestations might still be around). Perhaps more obviously, a psychological kind like *emotion* is inherent in minds (whether human or other), since an emotion is by its nature a mental state, and it doesn't seem to make sense to think of emotions that are not features of minds, no matter how different they may be from human minds. Given that the involvement of minds in psychological and social kinds is so fundamental, it isn't possible to say that the mind-dependence of such kinds is innocuous or superficial. Does that mean that we cannot be realists about such kinds? Rather than deny the reality of such kinds, it would be more reasonable to conclude that mind-dependence does not undermine realism about kinds. After all, the mind is a real phenomenon, which has evolved in at least one corner of the universe, so dependence on the mind should not be seen as a threat to realism. At least when it comes to kinds, there is no reason to regard mind-dependence and realism to be incompatible.

But before concluding definitively that mind-dependence is not inimical to realism about some kinds, it might be worth considering whether there are certain types of mind-dependence that might threaten realism. One of the most likely candidates is dependence on certain products of the mind, namely concepts, categories, or theories. It might seem as though kinds that are dependent in some ways on our theories can be fashioned by us at will and can therefore be freely manipulated by our thoughts. As such, they resemble fictional entities that exist only in our accounts of them, like Sherlock Holmes or Wonder Woman. But things aren't so simple, because many social kinds are indeed amenable to such manipulation by our categories and theories. As extensively demonstrated by Hacking (2006), many social kinds are "interactive" in this way and alter in response to our thoughts and theorizing. Consider the case of the social phenomenon of *child abuse*, which was labeled as such relatively recently, even though harm to children by caregivers has likely always been a feature of human societies. Hacking (1991b, 254) argues that in this and many other cases "people are affected by what we call them and, more importantly, by the available classifications within which they can describe their own actions and make their own constrained choices." This means that when a certain practice, such as corporal punishment in schools, is classified as "child abuse," those who engage in it may alter their behaviors by ceasing to engage in it, doing it more covertly, or modifying their actions in some other ways. In due course, the reaction of people to being classified or categorized will alter the practice itself, changing its nature, and perhaps leading us to revise our

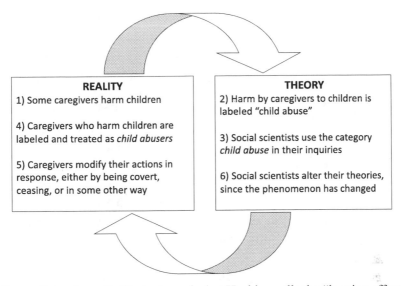

Figure 6 A schematic illustration of what Hacking calls the "looping effect" when it comes to some social kinds, such as *child abuse*, showing the way in which the category and the kind interact.

theories about it, or our very categories. What Hacking calls the "looping effect" is precisely this cycle whereby our categories and theories influence social actors in certain ways, who then alter their attitudes and behaviors, causing the social phenomenon itself to change, which in turn leads us to modify our categories and theories about that phenomenon. This cycle can be reiterated repeatedly (see Figure 6). Given that some social kinds are malleable in this way and alter in response to our theorizing about them, does that rule them out as real kinds?

There are a couple of things to notice about this feature of some social and psychological kinds before addressing the question of realism. First, Hacking sometimes writes as though looping requires awareness of the classifications on the part of those being classified. But this is not a constant feature of the phenomena of looping effects and interactive kinds. There surely needs to be awareness on the part of *someone* in society for looping to take place, but the people being classified need not themselves be aware of the classifications. For example, if a group of children are classified as having Attention Deficit Hyperactivity Disorder (ADHD), they might be treated in a certain way based on the classification, and that treatment may lead to alterations in their behavior, which may in turn alter the condition itself. But all this may occur without actual awareness of the classification on the part of the children themselves, though there is awareness, of course, on the part of some in society, such as healthcare workers, parents, and teachers (Khalidi 2010). The second thing to note about looping effects is that they may not be restricted to social and psychological kinds. As various commentators on Hacking's work have observed, looping effects can pertain to various biological phenomena that have social importance or significance. Bogen (1988) has pointed out that classifying cannabis plants as *illegal drugs* influences their properties and manner of cultivation. Similarly, Haslanger (1995) observes that categorizing animals as *food* has an impact on the size, behavior, and distribution of some animal populations. This may even lead to modifications in the genotypes and phenotypes of certain animals due to artificial selection. In fact, artificial selection is a process whereby humans, armed with categories and conceptions, alter, sometimes over many generations, certain strains of animals and plants, thus modifying their genotypes and phenotypes. Indeed, artificial selection can result in creating whole new kinds, like the domestic dog, *Canis familiaris* (Cooper 2004; Khalidi 2010). In these and other cases, our categories can modify the kinds that exist in the world – and not just in the psychological and social worlds but in the biological world as well.

Having taken a closer look at the phenomenon of interactive kinds, which are mind-dependent in ways that might appear at first sight to impugn their reality, we should be in a better position to address the realism question. The reason that kinds that are responsive to our categorizations may be considered unreal is that they are in some sense under our control and can be modified by us at will, much as novelists or playwrights can alter the characters in their fictional works while they write them. But in the examples mentioned, the modifications that occur to interactive kinds are not completely under our control. Our theorizing and collective actions can instigate changes in the world, which can end up modifying the kinds very significantly, including psychological, social, and biological kinds. But this does not mean that they are entirely at our behest or mere figments of our imagination. Whether it is *child abuse* or *C. familiaris*, these phenomena are not simply invented by us. In fact, such kinds blur the boundary between what is invented and what is discovered. But this should not be surprising given that, especially in the social world, human minds can give rise to phenomena like speculative bubbles that can have devastating effects on an economy, or declare wars that lead to widescale death and destruction, or harbor prejudices that result in oppression, exploitation, and enslavement. A better way of understanding what makes interactive kinds real can be related back to the causal theory of kinds described in Section 2.4. As long as there are groups of entities that share causal properties and enter into the same or similar causal regularities, we can say that they belong to real kinds. When it comes to *child abuse* and *money*, as well as *cannabis* and *C. familiaris*, they seem to be good candidates for being real kinds by that criterion, though the causes at issue are in part mental ones. The fact that they are mind-dependent, indeed interactively so, should not undermine the case for kindhood. Even though they may change in response to our categories and theories, that alone should not prevent us from taking a realist stance toward them.[28]

In earlier sections I questioned claims by some philosophers that natural kinds are generally constructed or "cocreated" by investigators. Instead, I defended a realist account of kinds as opposed to a quasi- or semi-realist attitude. But is that stance consistent with the acceptance of the mind-dependent and interactive kinds that have been the focus of this section? The semi-realist accounts mentioned earlier held kinds to be mind-dependent across the board, not just in some cases. For example, Boyd considers his account of natural kinds to be "constructionist" or "mind-dependent," and he writes: "on the accommodationist conception and on lots of others, natural kinds and their definitions are discipline-or-practice relative and are thus not 'mind independent'" (Boyd 2021, 2889). Similarly, Reydon (2016) puts forward a "cocreation" model of natural kinds, whereby kinds are codetermined by nature *and* investigators working in concert. He defends the position by citing different criteria for classifying genes, which he claims reflect different investigative contexts and are hence mind-dependent. These proposals take kinds *generally* to be a product of the mind and the world, because of the fact that human inquirers impose their own capacities, purposes, and interests when investigating the world, not just in the specific cases discussed in this section. This is usually put forward as a fundamental metaphysical claim about the nature of kinds. However, what is at issue here are the ways in which some real kinds depend on the mind and others don't. The interactive psychological, social, and biological kinds considered here are mind-dependent and interact with our mental states for reasons that go beyond the simple matter of investigating the world. Though they change in response to our categories and theories about them, it is not mere inquiry that cocreates them or delimits their boundaries, but rather interaction of a specific sort. The feedback loops discussed by Hacking and others causally impact these kinds because of the interventions that humans make into certain domains, effecting changes in those domains, whether deliberately or not. That is why acceptance of some mind-dependent kinds is not equivalent to adopting a semi-realist position toward all kinds or regarding all kinds as products of the

---

[28] Hacking (1995; 2006) also thinks that looping effects create epistemological challenges, since the kinds in question are "moving targets" and will tend to evade inquirers' attempts to pin them down. This epistemological issue will not be further pursued here, but see Mallon (2003), for a response. It should also be mentioned that Hacking is not a realist about kinds and considers them to be mind-dependent in ways that go beyond interactivity.

human mind. Considering all kinds to be mind-dependent obscures the significant fact that in some cases (e.g. *C. familiaris*, *ADHD*, *child abuse*), dependence on the mind alters the causal properties of kinds, and in other cases our psychological and social capacities causally sustain some kinds (e.g. *emotion, money*).

Finally, it's worth noting that mind-dependence and looping effects are not just associated with human minds. There are phenomena in the world that are dependent on *nonhuman* minds, and some kinds are modified or come into existence as a result of looping effects that pertain to the minds of other animals. Consider sexual selection in nonhuman animals. In some cases, the desire of certain individual organisms for mates with specific features is partly responsible for creating mates with those features. The properties of the peacock's tail are fashioned over generations by the desires and preferences of peahens, and the changes that occur to the tail modify those desires and preferences in turn. This is also true of some aspects of natural, not just sexual, selection. For example, in mimicry, members of animal or plant species typically evolve in such a way as to imitate the characteristics of other species in order to deceive the perceptual and cognitive capacities of members of yet other species. The mental capacities of some animal species are therefore instrumental in the causal process that leads to the selection of certain features rather than others. Those features are dependent on the minds of other creatures and go on to influence them accordingly. This sometimes leads to what has been called the Red Queen effect, whereby one species evolves to evade another, while the other evolves to overcome those defenses, and so on, resulting in an arms race.[29] What's important in this context is that this type of arms race can be mediated by the minds of the animals involved. But in all these cases, the products of the mind are types of entities with causal powers of their own, which is what makes them candidates for being real kinds, in line with the simple causal theory outlined in Section 2.4.

4 Applications: Kinds across the Sciences

In this section some of the conclusions reached about the metaphysics of kinds, the nature of kinds, and the various kinds of kinds will be put to work in analyzing a few case studies drawn from a diverse group of sciences. The aim is to show how these conclusions can be supported in specific cases and to derive further insights from these case studies about the nature of real kinds. Two of these cases have received some previous attention from philosophers of science, so there is a good body of work to build on. These are all cases that have

---

[29] The name "Red Queen effect" is a reference to Lewis Carroll's *Through the Looking-Glass*, in which the Red Queen explains to Alice that in Looking-Glass Land, one has to keep running just to stay in place (Van Valen 1973).

also garnered some attention outside of scientific and philosophical circles, instances in which questions of classification have been explicitly posed and debated in public forums, and even explicitly decided by bodies of experts. In Section 4.1, I will take a look at the category *planet* in astronomy, particularly the controversy surrounding whether to consider Pluto a planet. Then, in Section 4.2, I will consider the category *pandemic* in epidemiology, asking whether it is a good candidate for being a real kind of event or process. Finally, in Section 4.3, I will take up the category of *autism* (or people with autism) in psychiatry, which has recently been thought to be a spectrum rather than a single condition.

4.1 Planets

Every schoolchild now knows that Pluto was once considered a planet but that it isn't any longer. This raises the question as to what a planet is, whether the category *planet* corresponds to a real kind, and if so, on what grounds. At first glance, it might seem obvious that a planet is a large, approximately spherical object orbiting a star. But is there a way of nonarbitrarily specifying how large or massive the object must be? And is there a principled reason for requiring the object to be spherical, or for that matter, to be in orbit? And are the answers to these questions such that the category would correspond to a real kind?

At the beginning of the twenty-first century, there was widespread agreement among scientists (as well as among the lay public) that our solar system contained nine planets orbiting the sun. Pluto was the last of these bodies to be included among the planets and its existence was predicted at the beginning of the twentieth century on the basis of perturbations in the orbits of Uranus and Neptune. A few decades later, in 1930, a planet was observed in the predicted position, but it was later found to be too small to account for the observed perturbations. Nevertheless, Pluto was accepted as a planet, and so matters stood for decades.

In the 1990s, astronomers observed more Pluto-like objects in the Kuiper Belt, the region of the solar system beyond Neptune. In particular, in 2003, a Kuiper Belt Object (KBO) was discovered that was 27 percent more massive than Pluto, which was later named Eris (Magnus 2012, 76). This raised the question as to whether Eris and other such objects should also be considered planets, or whether they should all be disqualified. Why not admit all such KBOs as planets? One complication was that the orbits of some of these objects were not close to the ecliptic like those of other planets. Pluto itself was inclined 17° to the ecliptic, while Eris was inclined a full 44°. (By contrast, the eight other planets orbit more or less in the same plane, the most inclined being

Mercury at just 7° inclination.) Another complication was that there were objects in the asteroid belt (between Mars and Jupiter), like Ceres, that also seemed to have a strong claim to being planets. Ceres is roughly spherical and also orbits the sun, as the largest object in the asteroid belt. It was originally considered a planet in the early nineteenth century but reclassified as an asteroid in 1850. So should one consider such objects planets as well?

The issue was debated and eventually decided in 2006 at a meeting of the International Astronomical Union (IAU). The IAU defined "planet" as follows:

(1) A planet is a celestial body that
   (a) is in orbit around the Sun,
   (b) has sufficient mass for its self-gravity to overcome rigid body forces so that it assumes a hydrostatic equilibrium (nearly round) shape, and
   (c) has cleared the neighborhood around its orbit. (International Astronomical Union 2006)

This definition effectively excluded Pluto, as well as other contenders like Eris and Ceres, admitting only the eight planets that had been observed before the discovery of Pluto. But is this just an arbitrary set of criteria, or is there some principled reason for classifying some celestial objects on this basis? In other words, does this definition delineate a real kind?

Magnus (2012, 83) points out that criteria (b) and (c) are not independent, since they are both a result of having sufficient mass, and (c) implies (b). Accordingly, he rephrases the definition as follows: "*A planet is an object which is not itself a star but which is massive enough to dominate its orbit around a star*" (Magnus 2012, 83; original emphasis). In fact, there is a causal relationship between these criteria; they are not just an arbitrary collection of properties. An object of sufficient mass will be such as to dominate its orbit, since other, less massive objects will either be pulled into them or flung away from them. But any such object is likely to have a gravitational pull that will resolve it into a nearly spherical shape. Objects like Pluto, Eris, and Ceres are not massive enough to dominate their orbits in this way (though they are massive enough to be nearly spherical). Thus, the more funda-mental causal property of a planet is having a sufficient mass to clear its orbit. But one might still ask whether there is a principled reason for requiring planets to exceed a certain mass threshold so as to clear their orbit. Does this complex causal property correspond to a real kind?

As Bokulich (2014, 473–474) explains, the property of having a sufficient mass to clear its orbit is one that is significantly linked to the manner of formation of these objects in the first place:

An interstellar cloud of gas and dust initially collapses under gravitational attraction to form a star. Because the nebula is rotating the remaining gas and dust form a flat pancake-like disk rotating around the star. Accretion is the process by which these small particles collide and stick together forming a number of small planetesimals; the gravitational force of the larger planetesimals is then able to draw other smaller planetesimals to it eventually becoming large enough to form a planet. The larger a planetesimal gets, the more quickly and effectively it can gather even more material to it through gravitational attraction.

This physical process of accretion tends to result in a number of massive bodies in nonintersecting orbits that do not collide with one another. Hence, one can think of the kind *planet* as an etiological one that is related to the physical process of accretion. A planet is an object that has been formed in a certain way and it will tend to have a sufficient mass to clear its orbit because of its manner of formation. Moreover, having such a mass is also sufficient for it to achieve a hydrostatic equilibrium and assume a roughly spherical shape. When viewed in this way, it is clear that we have a cluster of causally linked properties rather than an arbitrary assortment of features. This causal network also accounts for the explanatory and predictive value of the category *planet*, thus reinforcing the claim that it corresponds to a real kind. As Magnus (2012, 84) observes, "the orbit-domination of planets allows us to explain facts about the formation, development, and present configuration of the solar system." Even though the IAU proposal is put in terms of a definition consisting of a set of necessary and sufficient conditions, it can be thought of as identifying a key property of planets that tends to be causally linked to other properties. Rather than a set of necessary and sufficient conditions, what we have is a causal clustering, in line with the simple causal theory (SCT) of real kinds presented in Section 2.4. Attending to the causal relationships between the properties of planets enables us to see that the kind *planet* is not a mere stable cluster of properties (along the lines of the stable property cluster or SPC theory mentioned in Section 2.4). Yet, there is no causal mechanism that keeps the properties of planets in homeostasis, as a strict version of the HPC theory would require (see Section 2.3).

Before concluding definitively that *planet* is a real kind in astronomy, as defined by the IAU, one aspect of the IAU definition (and Magnus' simplified version) that has not been addressed is the restriction to objects orbiting around stars. After all, the accretion process would seem to apply to sufficiently massive objects orbiting nonstars, say other planets. Indeed, the formation of moons can sometimes conform to the same physical process, so why not consider moons that have formed by accretion, dominate their orbit, and have a spherical shape to be planets orbiting other planets? While this process can also occur around planets, it is typically only stars that have sufficient mass to sustain a population of planets as characterized earlier. Moreover, moons are often the result of other physical processes, such as the capture of a wandering object by a planet when it approaches near enough, or a collision between the planet and some other object (as is thought to be the case for our moon). Thus, a planetary system formed by accretion pertains typically to stars and can be used to justify this restriction, though this is a question that would seem to warrant further discussion. It may be necessary to distinguish *solar* planets from *planetary* planets. This leads to a different question about the IAU definition, namely why it only pertains to our solar system. The IAU definition does not cover "exoplanets," planets associated with other stars. But given the generality of the causal process of accretion described earlier, it's possible to extend it to other systems in our galaxy and beyond (cf. Bokulich 2014, 484–485). There doesn't seem to be a principled reason to restrict the term to our solar system.

It is worth underlining that the cluster of properties associated with the kind *planet* includes relational and historical (etiological) properties as well as intrinsic ones (cf. Bokulich 2014, 480–481). This is entirely consistent with the defense of etiological and functional kinds in the previous section (see Sections 3.2 and 3.3). Despite the fact that some astronomers who opposed the IAU characterization of *planet* did so on the grounds that it cites relational rather than intrinsic properties (see Bokulich 2014, 469–470; Slater 2017, 6–7), the considerations cited earlier seem to vindicate the inclusion of such properties even in the domain of the physical sciences.

4.2 Pandemics

For obvious reasons, over the past few years there has been heightened awareness of epidemiological terms like "pandemic," "epidemic," "morbidity," and "zoonosis." Do such terms, particularly "pandemic," refer to real kinds in the epidemiological domain? Are there such things as pandemics with a stable set of properties or features that we can generalize about, or is each pandemic unique? In short, is *pandemic* a real kind with a cluster of causal properties that are related in certain ways?

The definition of a pandemic has long been in question among epidemiologists, and the difference between a pandemic and an epidemic has been the subject of some debate. There is a rough understanding that a pandemic is a large or geographically widespread epidemic, but geographic spread is clearly a vague notion, and this characterization also just raises the question of what an epidemic is. In one of the most systematic attempts to address the question, Morens, Folkers, and Fauci (2009) tried to come up with a set of features that characterize pandemics in general. By considering a range of diseases commonly said to be pandemic, they emerged with eight characteristics that are frequently associated with pandemics:

1) *Wide geographic extension*: "extend over large geographic areas"
2) *Disease movement*: "spread via transmission that can be traced from place to place"
3) *High attack rates and explosiveness*: have high rates of incidence among a population and "multiple cases appearing within a short time"
4) *Minimal population immunity*: affect populations that have low rates of immunity
5) *Novelty*: "are new, or at least associated with novel variants of existing organisms"
6) *Infectiousness*[30]
7) *Contagiousness*: are contagious from person to person
8) *Severity*: are severe or fatal diseases (Morens, Folkers, and Fauci 2009, 1019–1020)

There are a couple of important shortcomings in this proposal and, indeed, in the entire approach. First, the aim of these researchers is apparently mainly descriptive. They are attempting to determine how the term "pandemic" has usually been used in the scientific community, not to make a recommendation as to how it should be used, or indeed, whether it should be retained at all. For instance, they write: "The term pandemic has less commonly been used [in scientific discussions] to describe presumably noninfectious diseases, such as obesity" (Morens, Folkers, and Fauci 2009, 1019), without attempting to justify why it has been restricted in this way among the scientific community. Second, the authors admit that several of their criteria are either "relative" (e.g. *minimal population immunity, novelty*) or not consistently applied to pandemics (e.g. *high attack rates, infectiousness*). Perhaps for this reason, they conclude, apparently reluctantly, that "[t]here seems to be only 1 invariable common denominator: wide spread geographic extension" (Morens, Folkers, and Fauci 2009, 1020). But as first seen in Section 1.4, a single property cannot be considered to correspond to a real kind, since kinds are characterized by a cluster of properties, or at the very least, one property that is reliably linked to another. Moreover, on a causal account of real kinds, the link between the properties needs to be causal.[31]

---

[30] While the authors concede that the term "pandemic" has sometimes been used to describe diseases that are noninfectious, such as obesity, they maintain that the scientific community applies the term to infectious diseases.

[31] Arguably, *pandemic* might not be considered a real kind on any of the theories considered in Section 2. It does not seem to conform to an essentialist picture, since no other properties flow from the property of widespread geographic extension. It cannot be captured by the HPC theory, for the lack of a mechanism that keeps a cluster of properties in homeostasis. And it does not even qualify according to an epistemic theory like the SPC view, since there does not seem to be a stable cluster of properties that recur regularly.

But perhaps we can remedy the situation by taking a closer look at the properties that these researchers identify and trying to determine how they might be related. If we focus on the property of widespread geographic extension, that does not seem to be a cause of any of the other properties listed. If anything, it may be an effect of some of the others, such as high attack rates, novelty, and contagiousness. But it is not clear which of these properties or clusters of properties should be considered the primary or core properties of pandemics. Moreover, the link to widespread geographic extension, and indeed the precise extent of spread in question, are unclear. This is not to say that one could not come up with a causal model of pandemics that would show that these properties, or some subset of them, are causally related in such a way that they might tend to cluster together, albeit loosely. But as things now stand in epidemiology, there does not seem to be even a rough causal model of pandemics.

This conjecture is confirmed by some standard reference works on the subject. The *Dictionary of Epidemiology* published by the International Epidemiological Association defines a pandemic as follows: "An epidemic occurring over a very wide area, crossing international boundaries, and usually affecting a large number of people" (Porta 2014, 209). Moreover, the Centers for Disease Control in the United States asserts in one of its official publications, which is an introductory textbook on epidemiology: "Pandemic refers to an epidemic that has spread over several countries or continents, usually affecting a large number of people" (Centers for Disease Control and Prevention 2012, 503). It seems clear from these texts that the only property that is generally accepted as being associated with pandemics is that they are epidemics that are widely spread geographically. Even if we ignore the vagueness in that characterization, unless the property of geographic spread can be found to be associated with other properties, it cannot be said to identify a real kind.

It might be objected that since epidemiology is not a purely biological domain of inquiry, but rather involves social and political factors as well, the regular effects of pandemics should be sought in the social domain. But there don't seem to be any clear social consequences of pandemics that might delineate them as members of a real kind. A recent monograph on pandemics throughout history considers seven diseases that have had wide geographic distribution: plague, smallpox, malaria, cholera, tuberculosis, influenza, and HIV/AIDS, some of them having had such a distribution at more than one period in human history (McMillen 2016). But each of these diseases had a distinctive trajectory with no stable or regular features attested in every case or even in a substantial number of cases. This pertains to everything from transmission rates, severity, and morbidity, to socioeconomic effects and political consequences. To be sure, there are some intriguing resonances in a few cases. For example, the Black Death in Europe in the fourteenth century led to a decrease in population, which in turn led to labor shortages and higher wages combined with inflation (McMillen 2016, 12). These social consequences seem to foreshadow some of the effects of the COVID-19 pandemic in the early twenty-first century, but the resemblance between the two cases is merely skin-deep. In the former case, these social consequences resulted from the fact that up to 60 percent of the population of Europe died, whereas in the latter case, labor shortages and higher wages in many countries have not been the result of significant depopulation.

The other glaring problem with the commonly accepted definition of pandemics, which was briefly alluded to earlier, is that they are supposed to be *epidemics* that are geographically widespread. This suggests that *pandemic* is a subkind of *epidemic*, and it raises the question of what epidemics are and whether they are real kinds in their own right. We can't tackle this question in any detail here, but there is also reason for caution when it comes to the real-kind status of epidemics. The *Dictionary of Epidemiology* quoted earlier defines an epidemic as follows: "The occurrence in a community or region of cases of an illness, specific health-related behavior, or other health-related events clearly in excess of normal expectancy" (Porta 2014, 93).[32] The final clause "clearly in excess of normal expectancy" raises some red flags, since it seems to be relative to some standard of normal expectation among a group of observers, and neither the group nor the standard are specified. Moreover, the disjunction of three rather diverse phenomena in the definition ("illness, specific health-related behavior, or other health related events") appears quite heterogeneous and does not suggest any clustering of properties, whether medical or social. This is not to say that one couldn't specify the category of epidemics in such a way as to identify a cluster of properties, or indeed a causal clustering, but this does not appear to have been done as of yet. Since the definition of "pandemic" relies on the concept of epidemic, these problems cast further doubt on the real-kind status of *pandemic* (as well as *epidemic*).

The problems with defining the category *pandemic* and specifying its characteristic features have hardly gone unnoticed among researchers and specialists in epidemiology, and they have also led to controversy in public health and health policy circles. The World Health Organization has been criticized for altering its definition and for including arbitrary elements in its definitions:

---

[32] The rest of the entry reads: "The community or region and the period in which the cases occur must be specified precisely. The number of cases indicating the presence of an epidemic varies according to the agent, size, and type of population exposed; previous experience or lack of exposure to the disease; and time and place of occurrence. Epidemicity is thus relative to usual frequency of the disease in the same area, among the specified population, at the same season of the year."

> Statements from WHO ... suggest that pandemics are something inherently
> natural and obvious, out in the world and not the subject of human
> deliberation, debate and changing classificatory schemes. But what would
> and would not be declared a pandemic depends on a host of arbitrary factors
> such as who is doing the declaring and the criteria applied to make such
> a declaration. (Doshi 2011, 534)

This critique of the WHO definition brings out the lack of a nonarbitrary
characterization of pandemics. Perhaps as a result of such criticisms, the
WHO seems to have given up more recently on defining pandemics and has
been avoiding using the term in official declarations and publications, though
officials still use it in some of their statements: "International health organisa-
tions such as the WHO have not provided any formal definitions of the term
'pandemic', and the WHO no longer uses it as an official status of any outbreak"
(Singer, Thompson, and Bonsall 2021, 2547). This seems hardly surprising
given the evident problems with existing definitions. As we have already seen,
the set of properties proposed by Morens, Folkers, and Fauci (2009) seems to be
a mere list of features, several of which are underspecified or problematic, rather
than a cluster of causal properties that are related in certain ways.

So why should we have the category *pandemic* at all? Should it be discarded
altogether, since it does not seem to have genuine scientific value? The concept
may have some utility when organizing public health responses or mobilizing
public awareness. Doshi (2011) notes that in their responses to the H1N1 outbreak
of 2009, both the WHO and CDC had come up with definitions of "pandemic" in
some of their publications that referred to the *severity* of the disease involved, not
just its global spread. One researcher comments cynically on the short-lived
attempt to include severity in the definition of "pandemic" as follows:

> It is tempting to surmise that the complicated pandemic definitions used by
> the World Health Organization (WHO) and the Centers for Disease Control
> and Prevention of the United States of America involved severity in
> a deliberate attempt to garner political attention and financial support for
> pandemic preparedness. (Kelly 2011, 540)

For the time being, it is reasonable to conclude that *pandemic* is not a scientific
category that picks out a real kind of event or process,[33] but may rather be
a concept used to mobilize a concerted response to certain disease outbreaks.[34]

---

[33] Although pandemics are commonly described as "events" (McMillen 2016, 1), they are more
aptly thought of as processes, since they involve a sequence of spatiotemporally contiguous
events (cf. Section 1.3).

[34] Barnett (2011, 539) makes this mobilizing function of the category more explicit: "research
suggests that people are more likely to engage in desired protective behaviours in the face of
uncertain risk if they perceive the threat to be legitimately severe and relevant to them (and thus
motivating), and if they view the recommended intervention as efficacious. This would argue for

4.3 People with Autism

The overarching category of *mental disorders* or *psychiatric conditions* is laced with controversy and steeped in polemic. Addressing the real-kind status of any specific psychiatric condition is accordingly a very delicate matter.[35] But before approaching the question of whether *autism* is a real kind, it bears pointing out that the reality of a particular psychiatric condition could be independent of the reality of the superordinate category *psychiatric disorder*. There may be nothing in common to all the conditions that we consider to be psychiatric or mental disorders (cf. Zachar 2014), yet some of these conditions may be real psychiatric or mental kinds in their own right. An analogy might help: there might be nothing in common to all the species of animals that we consider pets, so *pet* may not be a real kind, but individual species (e.g. *dog*, *cat*, *goldfish*) may still be real kinds, regardless of whether the more general kind exists (cf. Murphy 2006, 98–99).

Autism is a category that has captured the public imagination and has been thrust into the headlines partly because of a recent classificatory controversy. The most widely accepted classification scheme of psychiatric conditions is the *Diagnostic and Statistical Manual of Mental Disorders* (DSM) published by the American Psychiatric Association. The fifth edition (DSM-5), which was published in 2013, made numerous changes to psychiatric taxonomy, including the removal of the separate category *Asperger syndrome* and its replacement with an overarching category of *Autism Spectrum Disorder* (ASD). As this label implies, autism[36] came to be conceived as a continuum ranging over those who have different degrees or levels of the disorder, with Asperger being a relatively mild form of the condition. The DSM-5 provides a set of criteria by which to diagnose ASD rather than attempting to specify its causal properties or give a theoretical account of it. Nevertheless, a look at these criteria can serve as a convenient starting point for characterizing ASD. The DSM-5 associates two overarching traits with ASD:

A. Persistent deficits in social communication and social interaction across multiple contexts . . .

B. Restricted, repetitive patterns of behavior, interests, or activities . . . (American Psychiatric Association 2013, 50–51).

---

severity as the main definitional predicate for pandemic declaration, rather than geography and virology."

[35] Philosophical interest in psychiatric taxonomy goes back at least to Hempel (1965); for further discussion of Hempel's view and an overview of philosophical discussions of classification with particular attention to psychiatry, see Mattu and Sullivan (2021).

[36] Apart from explicit discussion of the category mentioned in the DSM-5, I will use the term "autism" rather than "autism spectrum disorder," since it is not clear whether the condition should be conceived as a spectrum or as a disorder.

It goes on to further specify each of these components and provide further diagnostic criteria. When it comes to A, the criteria include deficits in social-emotional reciprocity, for example, reduced sharing of emotions and failure to initiate social interactions. They also include deficits in nonverbal communication, such as abnormalities in eye contact and body language. Moreover, the DSM-5 mentions deficits in maintaining relationships, such as making friends and absence of interest in peers. As for B, the criteria include stereotyped motor movements, inflexible adherence to routines, fixated interests that are abnormal in intensity, and hyperreactivity to sensory input. In some research on ASD, the distinctive traits of the condition are theorized as a trio rather than a duo: (i) communication deficits, (ii) social interaction difficulties, and (iii) repetitive patterns. This effectively splits the DSM-5's criterion A into two separate components (see e.g. Happé, Ronald, and Plomin 2006). Meanwhile, some psychiatrists conceive of autism as being rooted in one characteristic property, namely deficits in "theory of mind" (ToM) or the ability to understand the mental states of others, and it is sometimes regarded as "mindblindness" (Baron-Cohen, Leslie, and Frith 1985). This provides us with at least a preliminary characterization of the condition, paving the way for asking whether it might be a real kind.

A vast amount has been written about autism, including its redefinition as a spectrum and the absorption of Asperger syndrome, and it is impossible to try to tackle it in the scope of this section. But I will take a brief look at some of the issues involved and try to assess its real-kind status in light of some current research. To begin, it may be worth dispelling a concern about the way that the category is currently conceived in the DSM-5, as well as among many researchers and clinicians. Since autism is commonly theorized as a spectrum, that in itself may be thought to be incompatible with its being a real kind. But, at least on the simple causal theory of kinds being advocated in this Element, its being a spectrum doesn't necessarily preclude its being a kind. In the case of some properties that admit of degrees, there might be a threshold value at which certain effects follow, and these effects may consist in a cluster of properties that would correspond to a real kind. Both the causes and effects may lie along a spectrum, but as long as there is some point at which, or range in which, certain definite effects start to follow, this would be compatible with the existence of a real kind. However, if there is no such threshold and there is just continuous variation in the general population, then that would pose an obstacle to there being a distinct kind in play. Some researchers think that this latter situation is the one involved in autism:

> The distribution of such [autistic-like] traits supports a smooth continuum (at least at the behavioral level) between individuals meeting diagnostic criteria for ASD and individuals in the general population. Importantly, there is no evidence of a bimodal distribution, or "hump" at the extreme, separating clinical from nonclinical levels of difficulty. (Happé, Ronald, and Plomin 2006, 1218)

If they are right about this pattern of distribution, then it would be difficult to make the case that there is a distinct category of *autism* that corresponds to a real kind. But there does not seem to be a clear consensus on this question yet, and pending further research, we should explore other possibilities.

The main alternative possibility is that the traits implicated in autism are particularly pronounced in some members of the human species and that this results in a distinctive psychological type, with a characteristic cognitive, affective, and behavioral profile. For example, consider mindblindness. If it turns out that some people have certain deficits when it comes to their ability to understand other people's mental states, this trait may be causally responsible for at least some of the symptoms or traits associated with autism, including emotional and behavioral traits. This would mean that there is a causal network associated with autism whose core causal feature is mindblindness and whose effects include, say, reduced sharing of emotions and deficits in maintaining relationships, among other characteristic traits. Now, as many researchers have pointed out, it is unlikely that this single property can account for all the other traits commonly associated with autism. Boucher (2012, 238) writes: "Impaired ToM … has limited power to explain the full set of socio-emotional-communicative anomalies associated with ASD." For example, the restricted and repetitive behaviors mentioned in criterion B in the DSM-5 do not seem to follow causally from an impairment in ToM or mindreading ability. A more plausible alternative is that there is more than one causal property at the core of autism, such as the two main ones listed in the DSM-5 or the three posited by Happé, Ronald, and Plomin (2006). Moreover, it may be that when these three traits co-occur, they jointly generate the cognitive, affective, and behavioral profile associated with autism. On this "perfect storm" scenario, these three traits are independent but may just happen to be coinstantiated in certain individuals, and when they do, they will produce the full complement of psychological characteristics associated with autism. This would constitute a cluster of properties that are causally related: either the three basic traits are jointly responsible for the other characteristics of autism, or there are more complex relationships between the traits, whereby these traits cause some others, which in turn singly or jointly cause yet other traits associated with autism. It may also be that each of these core causal properties can occur without the others, and that when they do, they produce distinct sets of psychological traits (which may be partially overlapping). If so, then each of these clusters would constitute a somewhat different psychiatric kind, thereby splitting the kind *autism* into a number of different kinds. To mark this difference, we might choose to apply the term "autism" only to the condition whereby all three underlying traits co-occur, generating the "perfect storm," and to find different terms for the other conditions, which have only one or two of the underlying traits.

It may be objected here that the traits under discussion, such as the three mentioned by Happé, Ronald, and Plomin (2006), are mere symptoms rather than underlying causes. While some psychiatrists have put forward causal network models of psychopathology that involve only symptoms (e.g. Borsboom 2017; Borsboom and Cramer 2013), others suggest that causal models of psychiatric conditions can involve underlying causes or latent variables, not just symptoms (Bringmann and Eronen 2018). In the case of autism, each of the three traits mentioned earlier may have causal antecedents and may be the effects of psychological features that are causally upstream from them. Thus, social interaction difficulties may well be the result of impairments in ToM or mindreading, and this abnormality in understanding other people's minds may result in difficulties interacting with others in a social setting. Similarly, a propensity to engage in repetitive and restricted behaviors is likely to be the effect of a more basic psychological trait, and that cause might be thought of as one of the determining characteristics at the heart of autism. This would expand the causal cluster associated with autism to include not just cognitive, emotional, and behavioral symptoms, but their causal antecedents as well.

A different proposal has been made by Weiskopf (2017) who effectively proposes lumping rather than splitting these putative kinds. He suggests that ASD should be considered a "heterogeneous kind" even though he acknowledges that the very expression has "an aura of paradox" (2017, 182). Instead of conceiving each of the core properties mentioned earlier to generate its own kind (and the three together to give rise to yet another kind), he argues that the whole complex of features and traits should be considered to correspond to the kind *autism*. But if the underlying causal story is as posited here, then it seems that there would be little impetus to conceive of autism as comprising all these clusters. That is because those individuals who have only one of the underlying traits (e.g. communication deficits) may have little or nothing in common with individuals who have one of the other underlying traits (e.g. repetitive behaviors). This is acknowledged by Weiskopf, who states that on his proposed conception of autism, according to which the category consists in a "chain" rather than a cluster, "there may be no way to make inductive inferences from an arbitrary member to another" and that the individuals who are classified in this way "may have radically divergent capacities, experiences, and life prospects" (2017, 182). This would effectively render autism a disjunctive category that comprises distinct psychological profiles. However, heterogeneity within kinds seems to defeat the purpose of having kinds (recall Section 2.1). While cluster theories of kinds (like the homeostatic property cluster theory or the simple causal theory) allow properties to be loosely rather than strictly associated, there needs to be some homogeneity within the kind to have a kind at all.

More work is needed to determine whether these three traits are indeed causally responsible for the other characteristics associated with autism, as well as to distinguish individuals with all three traits from those who do not have all of them. There is also a need for research on the underlying psychological characteristics that may lie behind the three traits mentioned. As already indicated, the simple causal theory of kinds would suggest that people who have all the traits belong to a different psychiatric kind than those who have only some of them, since they consist in different causal clusters. If it turns out that these are indeed different kinds, then perhaps autism should not be theorized as a spectrum but should instead be "fractionated" into a set of different conditions (cf. Arnaud 2022, 12). Each cluster of causally connected traits can be thought of as a separate psychiatric kind, only one of which would be considered autism proper.

An obvious challenge needs to be considered to any account of autism as a real psychiatric kind. Many psychiatrists and philosophers of psychiatry would argue that what would vindicate autism as a real kind would be a connection to a neural mechanism (see e.g. Tsou 2021). So far, the discussion has focused entirely on psychological properties, including cognitive, affective, and behavioral characteristics, and has ignored any neural properties that might be associated with autism. However, as indicated in Section 3.2, many real kinds are functional rather than mechanistic or structural, hence the vindication of a real kind does not necessarily require the identification of a neural mechanism. As in other cases, causal clustering at the psychological level may be multiply realized by a diverse set of neural mechanisms (cf. Borsboom and Cramer 2019).

Finally, this entire discussion of autism as a psychiatric kind might be thought to ignore an obvious feature of psychiatric kinds, namely their value-ladenness. Psychiatric disorders, it might be said, are not just natural features of the world, they also have a normative dimension. A person diagnosed with autism does not have a neutral condition, like an atom that is characterized as ionized, or a plant described as an angiosperm, or even a patient diagnosed with cancer.

Psychiatric diagnoses carry certain alleged negative implications and are often stigmatized in the broader community. But this way of putting things elides an important distinction. There is no denying that categorization in the area of psychiatry can have serious ethical and social consequences for those classified, as well as for others (see e.g. Tekin 2014). But that is not to say that the categories themselves are or ought to be value-laden. To be sure, the super-ordinate category "psychiatric disorder" may be inherently evaluative and individuated in part by considerations of what is considered normal or accept-able behavior in a given society (and that is partly why it is not likely to correspond to a real kind, as I mentioned at the beginning of this section). But its subordinate categories need not all have this normative dimension. On a realist conception of kinds, the real kinds in a domain are identified when the inquiry is guided by epistemic values rather than moral, social, or other nonepistemic values (cf. Section 1.2).

4.4 Conclusion

In this section I have taken a brief look at three categories drawn from a diverse set of sciences: astronomy, epidemiology, and psychiatry. While I can't claim to have done justice to these three categories in the preceding sections, I hope that the treatment at least conveys a sense of the method of applying the philosoph-ical ideas presented in the preceding sections to actual case studies. In each case, the aim was to demonstrate how to go about determining whether a given scientific category corresponds to a real kind. This is sometimes done implicitly by scientists themselves when they attempt to define their categories or validate their constructs. But it is worth being more explicit about the enterprise, and the hope is that when we are, it will become clearer whether a specific category is a good candidate for being a real kind. Although I have presented a number of different views of what natural or real kinds are in previous sections, I have also openly defended a particular theory, namely the simple causal theory of kinds (see Section 2.4). Armed with this theory, I tried to emerge with a verdict on each of the case studies in this section. Based on the preceding discussion, it seems clear that *planet* is a real kind, since it corresponds to a cluster of properties that are causally linked. The kind is at once a functional kind (Section 3.2) and an etiological kind (Section 3.3), since it corresponds to a kind of astronomical body with a specific historical trajectory that features in certain causal regularities. By contrast, it is similarly clear that *pandemic* is not a real kind, precisely because it does not seem to exhibit the causal cluster-ing of properties and does not feature in causal regularities. At least as it is currently understood, it is a heterogeneous category that does not play a useful epistemic role in induction, explanation, or prediction. Meanwhile, *autism* is a promising candidate for a kind, even though as currently theorized, it may lump together more than one kind. Nevertheless, it does seem as though many of the symptoms associated with autism recur together and may be caused by a cluster of atypical underlying psychological characteristics.

References

Ahn, W., Kalish, C., Gelman, S. A. et al. (2001). Why essences are essential in the psychology of concepts. *Cognition* 82, 59–69.

American Psychiatric Association (2013). *Diagnostic and statistical manual of mental disorders*, 5th edition. American Psychiatric Association.

Anscombe, G. E. M. (1971). *Causality and determination: An inaugural lecture*. Cambridge University Press.

Armstrong, D. (1989). *Universals: An opinionated introduction*. Westview Press.

Arnaud, S. (2022). A social–emotional salience account of emotion recognition in autism: Moving beyond theory of mind. *Journal of Theoretical and Philosophical Psychology* 42(1), 3–18.

Barnett, D. J. (2011). Pandemic influenza and its definitional implications. *Bulletin of the World Health Organization* 89(7), 539.

Baron-Cohen, S., Leslie, A. M., and Frith, U. (1985). Does the autistic child have a "theory of mind"? *Cognition* 21(1), 37–46.

Bechtel, W. (2009). Looking down, around, and up: Mechanistic explanation in psychology. *Philosophical Psychology* 22(5), 543–564.

Bird, A. (2018). The metaphysics of natural kinds. *Synthese* 195(4), 1397–1426.

Bogen, J. (1988). Comments on "the sociology of knowledge about child abuse." *Noûs* 22(1), 65–66.

Bokulich, A. (2014). Pluto and the "Planet Problem": Folk concepts and natural kinds in astronomy. *Perspectives on Science* 22(4), 464–490.

Borsboom, D. (2017). A network theory of mental disorders. *World Psychiatry* 16(1), 5–13.

Borsboom, D., and Cramer, A. O. (2013). Network analysis: An integrative approach to the structure of psychopathology. *Annual Review of Clinical Psychology* 9(1), 91–121.

Borsboom, D., Cramer, A. O., and Kalis, A. (2019). Brain disorders? Not really: Why network structures block reductionism in psychopathology research. *Behavioral and Brain Sciences* 42, 1–63.

Boucher, J. (2012). Putting theory of mind in its place: Psychological explanations of the socio-emotional-communicative impairments in autistic spectrum disorder. *Autism* 16(3), 226–246.

Boyd, R. (1989). What realism implies and what it does not. *Dialectica* 43(1–2), 5–29.

Boyd, R. (1999). Homeostasis, species, and higher taxa. In R. A. Wilson (ed.), *Species: New interdisciplinary essays* (pp. 141–186). Massachusetts Institute of Technology Press.

Boyd, R. (2000). Kinds as the "workmanship of men": Realism, constructivism, and natural kinds. In J. Nida-Rümelin (ed.), *Rationality, realism, revision* (pp. 52–89). Walter de Gruyter.

Boyd, R. (2021). Rethinking natural kinds, reference and truth: Towards more correspondence with reality, not less. *Synthese* 198(12), 2863–2903.

Bringmann, L. F., and Eronen, M. I. (2018). Don't blame the model: Reconsidering the network approach to psychopathology. *Psychological Review* 125(4), 606–615.

Broad, C. D. (1920). The relation between induction and probability (Part II). *Mind* 29(113), 11–45.

Bursten, J. R. (2020). Smaller than a breadbox: Scale and natural kinds. *British Journal for the Philosophy of Science* 69(1), 1–23.

Centers for Disease Control and Prevention (2012). *Principles of epidemiology in public health practice*, 3rd edition. US Department of Health and Human Services.

Chakravartty, A. (2007). *A metaphysics for scientific realism: Knowing the unobservable*. Cambridge University Press.

Chen, L., DeVries, A. L., and Cheng, C. H. C. (1997). Convergent evolution of antifreeze glycoproteins in Antarctic notothenioid fish and Arctic cod. *Proceedings of the National Academy of Sciences* 94(8), 3817–3822.

Cooper, R. (2004). Why hacking is wrong about human kinds. *British Journal for the Philosophy of Science* 55(1), 73–85.

Crane, J. K. (2021). Two approaches to natural kinds. *Synthese* 199(5–6), 12177–12198.

Craver, C. F. (2007). *Explaining the brain*. Oxford University Press.

Craver, C. F. (2009). Mechanisms and natural kinds. *Philosophical Psychology* 22(5), 575–594.

Darden, L., and Maull, N. (1977). Interfield theories. *Philosophy of Science* 44(1), 43–64.

Doshi, P. (2011). The elusive definition of pandemic influenza. *Bulletin of the World Health Organization* 89(7), 532–538.

Dupré, J. (2002). Is "natural kind" a natural kind term? *The Monist* 85(1), 29–49.

Elder, C. (2004). *Real natures and familiar objects*. Massachusetts Institute of Technology Press.

Ellis, B. (2001). *Scientific essentialism*. Cambridge University Press.

Ereshefsky, M. (2001). *The poverty of the Linnaean hierarchy*. Cambridge University Press.

Ereshefsky, M., and Matthen, M. (2005). Taxonomy, polymorphism, and history: An introduction to population structure theory. *Philosophy of Science* 72(1), 1–21.

Ereshefsky, M., and Reydon, T. A. (2015). Scientific kinds. *Philosophical Studies* 172(4), 969–986.

Fodor, J. A. (1974). Special sciences (or: the disunity of science as a working hypothesis. *Synthese* 28(2), 97–115.

Franklin, F., and Franklin, C. L. (1888). Mill's natural kinds. *Mind* 13(49), 83–85.

Franklin-Hall, L. (2015). Natural kinds as categorical bottlenecks. *Philosophical Studies* 172(4), 925–948.

Ghiselin, M. T. (1974). A radical solution to the species problem. *Systematic Biology* 23(4), 536–544.

Glennan, S. (2017). *The new mechanical philosophy*. Oxford University Press.

Godman, M. (2021). *The epistemology and morality of human kinds*. Routledge.

Goodman, N. (1955/1983). *Fact, fiction, and forecast*. Harvard University Press.

Griffiths, P. E. (1993). Functional analysis and proper functions. *British Journal for the Philosophy of Science* 44(3), 409–422.

Griffiths, P. E. (1994). Cladistic classification and functional explanation. *Philosophy of Science* 61(2), 206–227.

Griffiths, P. E. (1999). Squaring the circle: Natural kinds with historical essences. In R. A. Wilson (ed.), *Species: New interdisciplinary essays* (pp. 209–228). Massachusetts Institute of Technology Press.

Hacking, I. (1991a). A tradition of natural kinds. *Philosophical Studies* 61(1–2), 109–126.

Hacking, I. (1991b). The making and molding of child abuse. *Critical Inquiry* 17(2), 253–288.

Hacking, I. (1995). The looping effects of human kinds. In D. Sperber, D. Premack, and A. J. Premack (eds.), *Causal cognition: A multidisciplinary debate* (pp. 351–394). Oxford University Press.

Hacking, I. (2006). Kinds of people: Moving targets. *Proceedings of the British Academy* 151, 285–318.

Hacking, I. (2007). Natural kinds: Rosy dawn, scholastic twilight. *Royal Institute of Philosophy Supplements* 61, 203–239.

Happé, F., Ronald, A., and Plomin, R. (2006). Time to give up on a single explanation for autism. *Nature Neuroscience* 9(10), 1218–1220.

Haslanger, S. (1995). Ontology and social construction. *Philosophical Topics* 23(2), 95–125.

Havstad, J. C. (2021). Complexity begets crosscutting, dooms hierarchy (another paper on natural kinds). *Synthese* 198(8), 7665–7696.

Hawley, K., and Bird, A. (2011). What are natural kinds? *Philosophical Perspectives* 25(1), 205–221.

Hempel, C. G. (1965). Fundamentals of taxonomy. In C. G. Hempel, *Aspects of scientific explanation and other essays in the philosophy of science* (pp. 137–154). Free Press.

Hull, D. (1978). A matter of individuality. *Philosophy of Science* 45(3), 335–360.

International Astronomical Union. (2006). Resolution B5: Definition of a planet in the solar system. www.iau.org/static/resolutions/Resolution_GA26-5-6.pdf.

Kahane, H. (1969). Thomason on natural kinds. *Noûs* 3(4), 409–412.

Keil, F. C. (2003). Folkscience: Coarse interpretations of a complex reality. *Trends in Cognitive Sciences* 7(8), 368–373.

Kelly, H. (2011). The classical definition of a pandemic is not elusive. *Bulletin of the World Health Organization* 89(7), 540–541.

Khalidi, M. A. (1993). Carving nature at the joints. *Philosophy of Science* 60(1), 100–113.

Khalidi, M. A. (1998). Natural kinds and crosscutting categories. *Journal of Philosophy* 95(1), 33–50.

Khalidi, M. A. (2009). How scientific is scientific essentialism? *Journal for General Philosophy of Science* 40(1), 85–101.

Khalidi, M. A. (2010). Interactive kinds. *British Journal for the Philosophy of Science* 61(2), 335–360.

Khalidi, M. A. (2013). *Natural categories and human kinds: Classification in the natural and social sciences*. Cambridge University Press.

Khalidi, M. A. (2017). Crosscutting psycho-neural taxonomies: The case of episodic memory. *Philosophical Explorations* 20(2), 191–208.

Khalidi, M. A. (2018). Natural kinds as nodes in causal networks. *Synthese* 195 (4), 1379–1396.

Khalidi, M. A. (2020). Are sexes natural kinds? In S. Dasgupta, R. Dotan, and B. Weslake (eds.), *Current controversies in philosophy of science* (pp. 163–176). Routledge.

Khalidi, M. A. (2021). Etiological kinds. *Philosophy of Science* 88(1), 1–21.

Kim, J. (1992). Multiple realization and the metaphysics of reduction. *Philosophy and Phenomenological Research* 52(1), 1–26.

Kitcher, P. (1992). The naturalists return. *Philosophical Review* 101(1), 53–114.

Kitcher, P. (2001). *Science, truth, and democracy*. Oxford University Press.

Kripke, S. (1972/1980). *Naming and necessity*. Harvard University Press.

Langton, R., and Lewis, D. (1998). Defining "intrinsic." *Philosophy and Phenomenological Research* 58(2), 333–345.

Lemeire, O. (2021). No purely epistemic theory can account for the naturalness of kinds. *Synthese* 198(12), 2907–2925.

Lewis, D. (1983). Extrinsic properties. *Philosophical Studies* 44(2), 197–200.

Ludwig, D. (2018). Letting go of "natural kind": Towards a multidimensional framework of non-arbitrary classification. *Philosophy of Science* 85(1), 31–52.

Machamer, P., Darden, L., and Craver, C. F. (2000). Thinking about mechanisms. *Philosophy of Science* 67(1), 1–25.

Magnus, P. D. (2012). *Scientific enquiry and natural kinds: From planets to mallards*. Palgrave Macmillan.

Magnus, P. D. (2013). No grist for Mill on natural kinds. *Journal for the History of Analytic Philosophy* 2(4), 1–15.

Magnus, P. D. (2014). NK ≠ HPC. *The Philosophical Quarterly* 64(256), 471–477.

Magnus, P. D. (2018). How to be a realist about natural kinds. *Disputatio: Philosophical Research Bulletin* 7(8), 1–11.

Mallon, R. (2003). Social construction, social roles, and stability. In F. Schmitt (ed.), *Socializing metaphysics: The nature of social reality* (pp. 327–354). Rowman & Littlefield.

Massimi, M. (2014). Natural kinds and naturalised Kantianism. *Noûs* 48(3), 416–449.

Mattu, J., and Sullivan, J. A. (2021). Classification, kinds, taxonomic stability and conceptual change. *Aggression and Violent Behavior* 59(1), 101477.

McMillen, C. W. (2016). *Pandemics: A very short introduction*. Oxford University Press.

Mill, J. S. (1843/1882). *A system of logic* (8th edition). Harper & Brothers.

Millikan, R. (1999). Historical kinds and the "special sciences." *Philosophical Studies* 95(1–2), 45–65.

Millikan, R. (2000). *On clear and confused ideas: An essay about substance concepts*. Cambridge University Press.

Millikan, R. (2005). Why (most) concepts aren't categories. In H. Cohen and C. Lefebvre (eds.), *Handbook of categorization in cognitive science* (pp. 305–316). Elsevier.

Mitchell, S. (2000). Dimensions of scientific law. *Philosophy of Science* 67(2), 242–265.

Morens, D. M., Folkers, G. K., and Fauci, A. S. (2009). What is a pandemic? *The Journal of Infectious Diseases* 200(7), 1018–1021.

Murphy, D. (2006). *Psychiatry in the scientific image*. Massachusetts Institute of Technology Press.

Page, M. (2021). The role of historical science in methodological actualism. *Philosophy of Science* 88(3), 461–482.

Papineau, D. (2010). Can any sciences be special? In C. Macdonald and G. Macdonald (eds.), *Emergence in mind* (pp. 179–197). Oxford University Press.

Peirce, C. S. (1901). "Kind." In J. M. Baldwin (ed.), *Dictionary of philosophy and psychology*. Macmillan. http://psychclassics.yorku.ca/Baldwin/ Dictionary/defs/K1defs.htm#Kind.

Porta, M. (2014). *A dictionary of epidemiology*. Oxford University Press.

Pöyhönen, S. (2016). Memory as a cognitive kind: Brains, remembering dyads, and exograms. In C. Kendig (ed.), *Natural kinds and classification in scientific practice* (pp. 145–156). Routledge.

Putnam, H. (1975). *Mathematics, matter, and method*. Cambridge University Press.

Quine, W. V. (1969). Natural kinds. In W. V. Quine, *Ontological relativity and other essays* (pp. 114–138). Columbia University Press.

Reydon, T. A. C. (2006). Generalizations and kinds in natural science: The case of species. *Studies in History and Philosophy of Biological and Biomedical Sciences* 37(2), 230–255.

Reydon, T. A. C. (2016). From a zooming-in model to a co-creation model: Towards a more dynamic account of classification and kinds. In C. Kendig (ed.), *Natural kinds and classification in scientific practice* (pp. 59–73). Routledge.

Ross, L. N. (2020). Causal concepts in biology: How pathways differ from mechanisms and why it matters. *British Journal for the Philosophy of Science* 72(1), 131–158.

Santana, C. (2019). Mineral misbehavior: Why mineralogists don't deal in natural kinds. *Foundations of Chemistry* 21(3), 333–343.

Schaffer, J. (2003). Is there a fundamental level? *Nous* 37(3), 498–517.

Singer, B. J., Thompson, R. N., and Bonsall, M. B. (2021). The effect of the definition of "pandemic" on quantitative assessments of infectious disease outbreak risk. *Scientific Reports* 11(1), 1–13.

Slater, M. H. (2015). Natural kindness. *British Journal for the Philosophy of Science* 66(2), 375–411.

Slater, M. H. (2017). Pluto and the platypus: An odd ball and an odd duck – on classificatory norms. *Studies in History and Philosophy of Science Part A* 61, 1–10.

Tahko, T. (2021). *Unity of science*. Cambridge University Press.

Tekin, Ş. (2014). Psychiatric taxonomy: At the crossroads of science and ethics. *Journal of Medical Ethics* 40(8), 513–514.

Thomason, R. H. (1969). Determinates and natural kinds. *Noûs* 3(1), 95–101.

Tobin, E. (2010). Crosscutting natural kinds and the hierarchy thesis. In H. Beebee and N. Sabbarton-Leary (eds.), *The semantics and metaphysics of natural kinds* (pp. 179–191). Routledge.

Tsou, J. Y. (2021). *Philosophy of psychiatry*. Cambridge University Press.

Van Valen, L. (1973). A new evolutionary law. *Evolutionary Theory* 1, 1–30.

Van Valen, L. (1976). Individualistic classes. *Philosophy of Science* 43(4), 539–541.

Venn, J. (1866/1888). *The logic of chance* (3rd edition). Macmillan.

Venn, J. (1889/1907). *The principles of empirical or inductive logic* (2nd edition). Macmillan.

Weiskopf, D. (2011a). The functional unity of special science kinds. *British Journal for the Philosophy of Science* 62(2), 233–258.

Weiskopf, D. (2011b). Models and mechanisms in psychological explanation. *Synthese* 183(3), 313–338.

Weiskopf, D. (2017). An ideal disorder? Autism as a psychiatric kind. *Philosophical Explorations* 20(2), 175–190.

Whewell, W. (1840/1847). *The philosophy of the inductive sciences* (2nd edition). John W. Parker.

Wilson, R. A., Barker, M. J., and Brigandt, I. (2007). When traditional essentialism fails: Biological natural kinds. *Philosophical Topics* 35(1/2), 189–215.

Wimsatt, W. C. (2007). *Re-engineering philosophy for limited beings: Piecewise approximations to reality.* Harvard University Press.

Zachar, P. (2014). *A metaphysics of psychopathology.* Massachusetts Institute of Technology Press.

Acknowledgments

I am very grateful to Tyler Olds and Maximiliana Rifkin for detailed comments on the draft manuscript and for discussion of some of the issues raised in this Element. The constructive criticisms of two incisive anonymous referees also led to numerous improvements. I am also grateful for a grant from the Social Sciences and Humanities Research Council (SSHRC) of Canada that enabled this publication to be open access.

Cambridge Elements ≡

The Philosophy of Science

Jacob Stegenga

University of Cambridge

Jacob Stegenga is a Reader in the Department of History and Philosophy of Science at the University of Cambridge. He has published widely on fundamental topics in reasoning and rationality and philosophical problems in medicine and biology. Prior to joining Cambridge he taught in the United States and Canada, and he received his PhD from the University of California San Diego.

About the Series

This series of Elements in Philosophy of Science provides an extensive overview of the themes, topics and debates which constitute the philosophy of science. Distinguished specialists provide an up-to-date summary of the results of current research on their topics, as well as offering their own take on those topics and drawing original conclusions.

Cambridge Elements ☰

The Philosophy of Science

Elements in the Series

Printed in the United States
by Baker & Taylor Publisher Services